BodyVoyage™

A Three-Dimensional Tour of a Real Human Body

BodyVoyage™

ALEXANDER TSIARAS

WARNER
BOOKS

A Time Warner Company

AUTHOR FOR ANATOMICAL TRAVELOGUE: Alexander Tsiaras

ART DIRECTOR FOR ANATOMICAL TRAVELOGUE: Tony Wurman

WRITER: Jeff Goldberg

ANATOMIST: Professor Philip Brandt

ADVISOR: Dr. Michael Vannier

3-D COVER: Andrew Joel and Chris Robin at Image Technology Inc.

HARDWARE:

Silicon Graphics Inc.: John Flynn

Sun Microsystems Inc.: Van Jepson

Apple Comupters Inc.: Robert Sparno

Olympus America Inc.: Precision Instuments Division: Dan Biondi

Sony New Technologies Inc.: Mitchell Cannold

Radius Inc.: Chuck Berger

SOFTWARE:

ISS/CieMed: Raghu Raghavan, Ph.D., Rakesh Mullick, Ph.D, H.T. Nguyen, Ph.D.,
    Chun Pong Yu, Ph.D., Meiyappan Solaiyappan, Ph.D.

Adobe Systems Inc.: John Warnock, Ph.D.

NAG Inc. (IRIS Explorer): Philip Harnden

Cemax: Rodica Schileru-Key

DATA:

The data used in Body Voyage was provided by the National Library of Medicine's
Visible Human Project. This data is provided on an interim basis and may be modi-
fied substantially by the National Library of Medicine in subsequent versions.
Anatomical Visualization Inc.: Victor Spitzer, Ph.D.

SPECIAL THANKS TO:

Larry Kirshbaum

Chee Pearlman

Evan Gsell (ballbuster)

BodyVoyage™

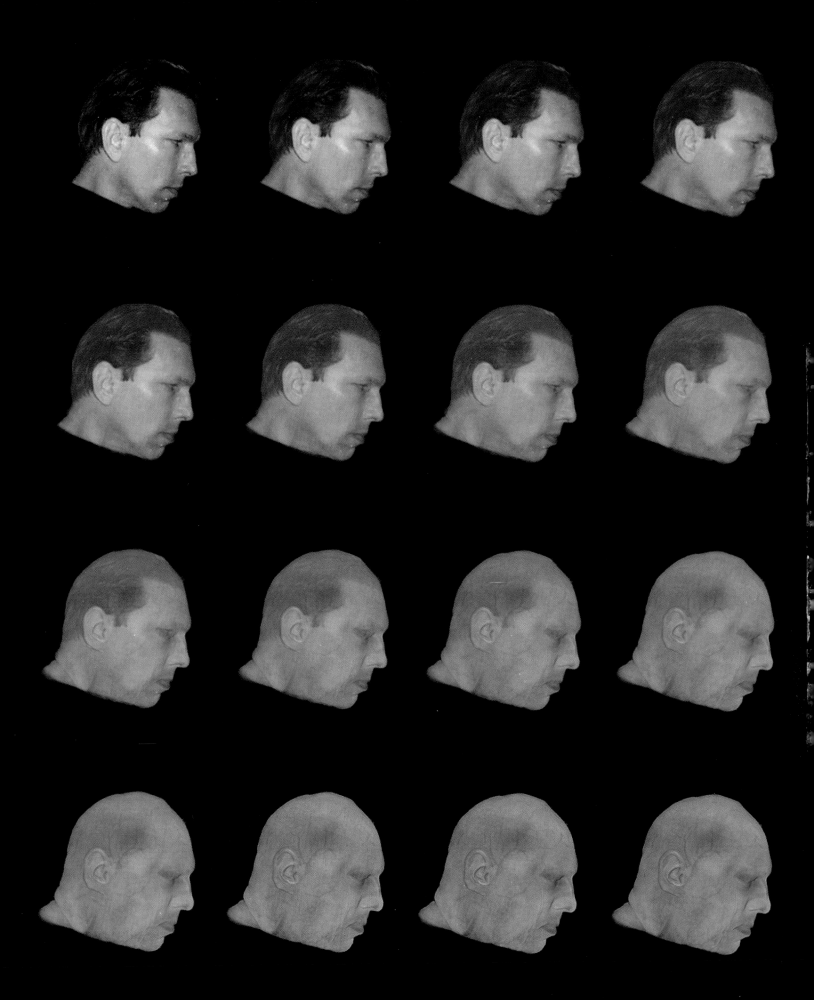

Introduction
by John Hockenberry

In the human chronicle of stepping back to get a better view, this book is a major event. We long for some complete grasp of our existence and yet contemporary civilization has subdivided time, space, soul, and body into smaller and diminishingly comprehensible units. We are drawn to think of our bodies as the product of inputs and the manifestation of outputs. Our health is a search for the solution to some increasingly complex equation that adds fiber, subtracts ozone, multiplies car exhaust, and divides too much fat by too little broccoli. Yet we resist such a reduction to find sanctuary here among these images.

I am inclined at the outset to say to the reader, viewer, or user: "Kindly check any preconceived notions about biology at the door, you may pick them up again when you leave (if you still think you need them)." For it is neither the meticulous detail, the selection of colors, the soothing soundtrack, nor the use of state-of-the-art computerized graphics that stays with you long after viewing *Body Voyage*. It hits you slowly in the pirouette of a familiar cylindrical geometry, rotating gently on the screen with a beauty that strikes harmonic chords on the body like smooth, Greek-sculpted marble. The images are tantalizing and peaceful, difficult to objectify. It requires effort to recall that what you are really looking at is a disembodied pelvis detached with a clean slice through the abdomen and the femurs, the interior of a skull, or a detached arm.

In a manner not immediately evident from the technology, designer Tsiaras has simultaneously sculpted and dissected an actual human body, achieving a sense of art and tranquillity worthy of a da Vinci. He has achieved it by using the virtual methodology of an ancient Egyptian embalmer or a digital Jeffrey Dahmer. Whether viewed on the page or in the three-dimensional moving pictures on CD-ROM, this is no idealized Cad-Cam rendering by an engineer. This is not an approximation of the human form. Neither is it an idealization like *Venus de Milo*, or Michaelangelo's *David*. Relatively little is abstract here.

Unlike renderings that make a whole body by connecting some key dots, each dot of printed ink, each glowing pixel in *Body Voyage*, maps to a specific place on an actual human body.

The body has lived at the same address since October 12, 1994: http://www.nlm.nih.gov. Before he took up residence on the Internet (in this precise sequence), Joseph Paul Jernigan was a burglar and a drifter, a convicted murderer, and death-row inmate who was executed with a massive dose of barbiturates, his cadaver frozen and meticulously sliced into 1,871 pieces. Those pieces were scanned and digitized into an archive that belongs to the National Institutes of Health. *Body Voyage* is rendered from this single set of data.

Given the elaborate trauma that preceded Jernigan's arrival in digital space, it is designer Tsiaras's great achievement that he shows us places inside the body without compromising its structure and symmetry. The individually sliced and scanned images from the Internet have been recombined with other medical data about Jernigan, enhanced with colors chosen by the artist, to produce what is called a multimodal rendering. This technique allows a three-dimensional image of the human exterior and interior to be presented with identical detail and vividness.

How the data in *Body Voyage* was acquired is in no sense apparent from watching the beautiful images rotating on our digital axes. Our point of view inside the skull, peering at the back side of the eye sockets, demands no gaping hole in the cranium for the insertion of camera and flashlight. We scan quickly around and see the bones that cradle the ears that form the hinges of the jaw, and there behind where we were just looking is the back of the skull and the channel for routing the base of the brain into the spine. It is all there unbroken. What was once revealed only through the cutting and removal of tissue is now seen intact, frozen, and functionless, a quiet cathedral, its congregation and clergy removed, but the sacred is there whispering and echoing through the miraculous structure. This is anatomy undisturbed by the intrusive dissector's instruments of exploration.

The science of anatomy is nearly unique among the medical disciplines in that it offers no therapeutic value in and of itself. But it is a crucial body of knowledge because anatomy is the measure of what we know, and perhaps more revealing, what we are permitted to know about the human organism. Images of the human body's interior have evolved down through time as radically as the map of the world. But the human body has never been as remote as the planetary frontier. It has always been, literally, right under our noses. But with our bodies explorable and available only in theory, humans have always had to cross moral wilderness to stake out truths in medical science rather than oceans, prairies, and mountain ranges. Not all of the places the explorers have found have been equally hospitable.

For much of world history, exploring the human body was heresy and dissecting for research a capital crime. In the 1700s the brisk black market trade in cadavers in London, Paris, and Germany built the first fortunes of modern medical science and directly confronted the church's notions of free will and the soul's role in healing. Yet the insights about life gained from letting go of the sanctity of the dead body have formed the basis, in our time, of the extraordinary success in medical healing of live bodies.

These images are a direct descendant of the documents of human moral transformation recorded in the serene details of the anatomical drawings of the Renaissance and Enlightenment with their archaic arrows and numbered annotations. From the woodcuts of Galen to the muscles and bones of da Vinci's sketches, from the plastic overlays of an encyclopedia and medical textbooks to the three-dimensional images on advanced computers, images of the human body are images of humanity as well. In contrast to *Body Voyage*, the detailed drawings of muscle and bone in the standard textbook of medical imagery, *Gray's Anatomy*, are best described as artfully detailed pictures of dissections. *Gray's Anatomy* would easily fit on a CD-ROM today. Yet even the visual data in *Gray's* is the least important part of its data set. *Gray's Anatomy* is not only pictures but also the collected record of medical choices and techniques over more than three centuries. All of the renderings of the human form down through history plot a path toward an understanding of the literal structures of life gained at an apparent cost to the reverence for the dead.

From a time when theologians confidently described the soul with surgical precision, and the internal structure and function of the body was merely guessed at (Aristotle once surmised the brain was a cooling system for the blood), we have arrived at a moment where the body has been mapped down to the molecular level. Today it is the soul we search for, guessing about it in the brash and uneasy manner of alchemists while we remain supremely confident of our abilities to heal and manipulate the flesh. Floating through the skull and heart and lungs of Alexander Tsiaras's world we see every detail of our knowledge about the body and our ability to measure and define its function. We see here a body utterly and vividly intact. Only the soul has been removed.

Explore this world and I think you will agree with me that whatever else is to be found in this collaboration between an executed criminal and an artist we are somehow deftly aided in the ancient search for the soul. It is the oddest of ironies. The atomized, digitized body of Joseph Paul Jernigan reconstituted and imbued with a mesmerizing beauty and realism is as good an argument for the tangibility of the soul as one can find in this cheerless age of cause and effect. Here artist Tsiaras gives us more to look at than ever before, separating the layers of flesh and fat and bone with cellular precision. If there is to be an Audubon of the human form, Tsiaras can make rightful claim to the title. But the digital features of *Body Voyage* make it much more flexible than a field guide's set of illustrations. *Body Voyage* might also eventually become a literal archive for recording future users' points and clicks, choices they might make in exploring the pictures and data. What will explorers in the distant future see here? As was the case hundreds of years ago, the future of medical science is no deterministic march of technology. It lies instead in what the next generation of explorers wants to know. The future lies in the questions they will ask, questions that cannot be imagined now. That is as it should be.

Joseph Paul Jernigan may have given his body to science, but watching his slowly rotating body in 3-D digital space, one feels that through the work of artist and designer Alexander Tsiaras, Jernigan has given it to humanity as well. With quiet examination these images come to life as surely as the characters of Bosch move on the surface of his paintings, as surely as stained glass angels soar in the cathedral at Rheims. Whatever tragic legacy Joseph Paul Jernigan left in life, in death he has found grace.

Not even Lazarus looked this good.

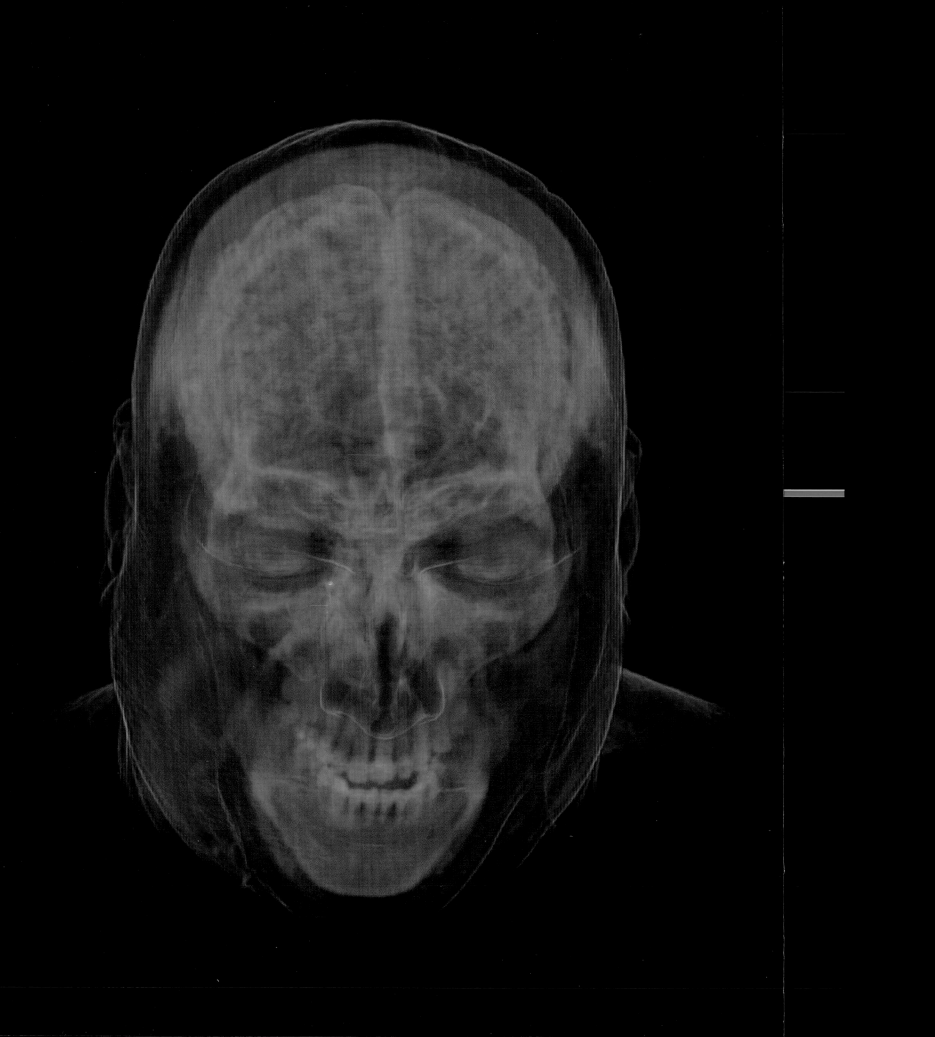

The Head

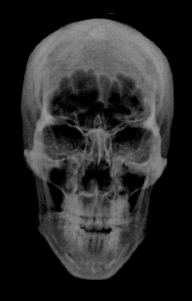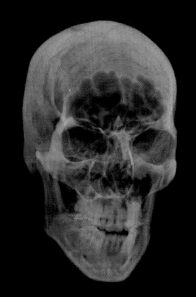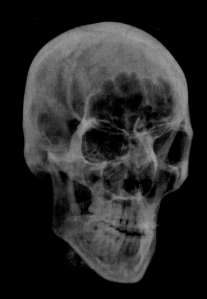

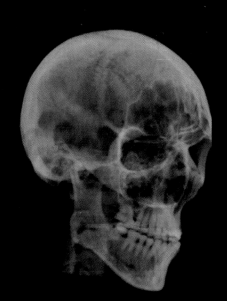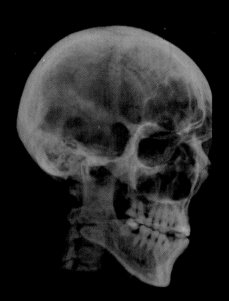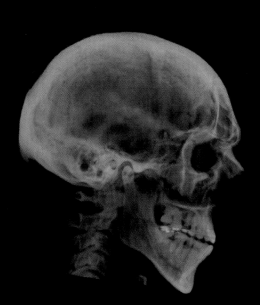

The human face, the calling card of our individual identity, is only a $1/100^{th}$-of-an-inch layer of skin cells, stretched over a skeletal framework of 14 facial bones. The greater mystery of who we are lies within it.

Millions of years of evolution have crafted the highly organized tissues and organs that allow us to see, hear, and smell the world around us; the flexible mask of muscles that can produce the hundreds of distinctive facial expressions that color all human interactions; and the uniquely designed structures of the throat and mouth that enable us to speak.

Probing deeper, beneath the protective yet vulnerable helmet of the skull, the appearance of structure gives way to chaotic complexity. Yet here, within the three-pound universe of the cells that govern our intellect, intuition, memories, perceptions, and emotions, the secrets of our true identity are hidden.

No two faces and no two brains, not even those of identical twins, are exactly alike. Scientists are intrigued by the possibility that subtle workings of the brain could distinguish a murderer from a saint, in the same way that the well-developed muscles that molded Joseph Paul Jernigan's bullish neck, the enormous sinus cavities that created his jutting brow, and other slight differences in skin, bone, and muscle contributed to his unique appearance.

The answers remain largely unknown, but it is clear that they are not skin-deep.

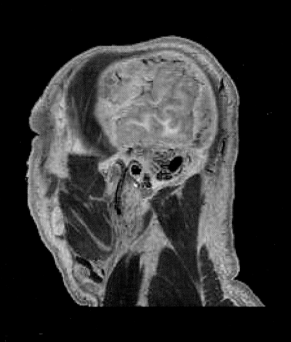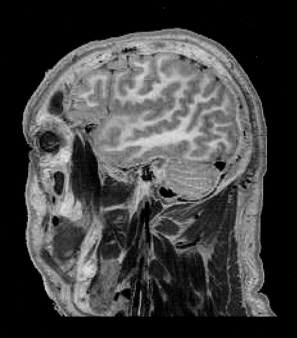
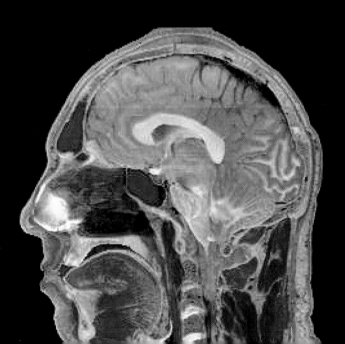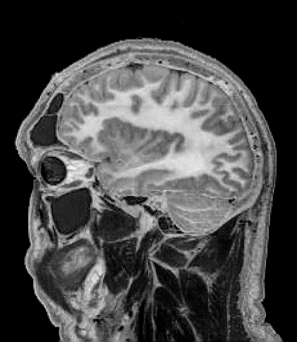

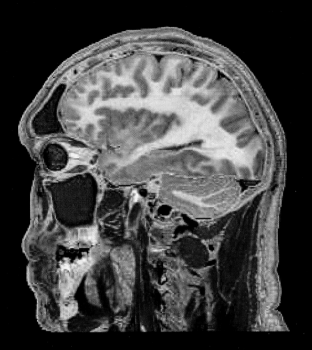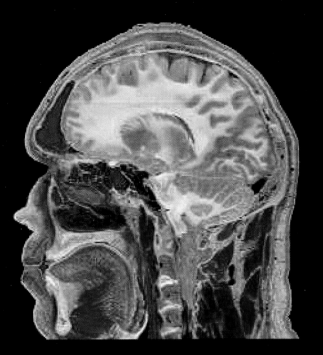
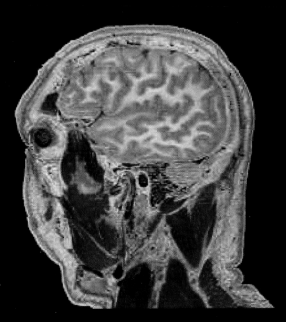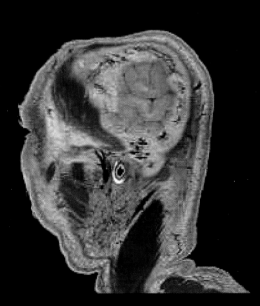

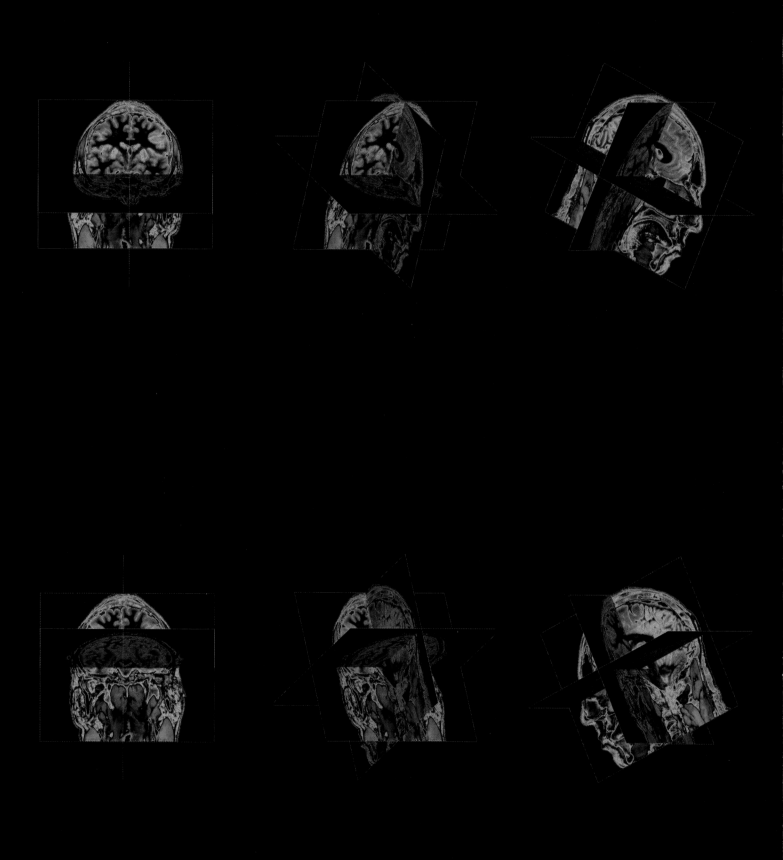

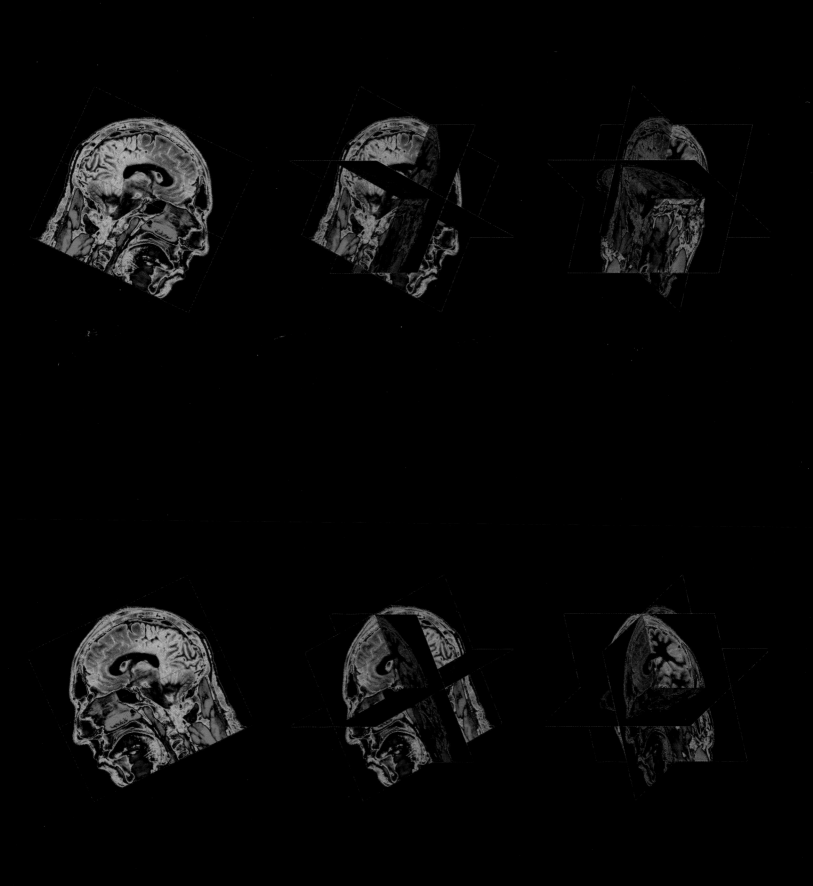

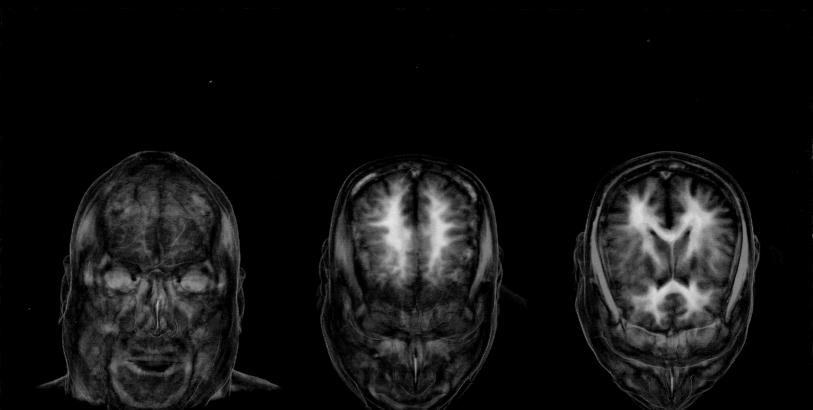

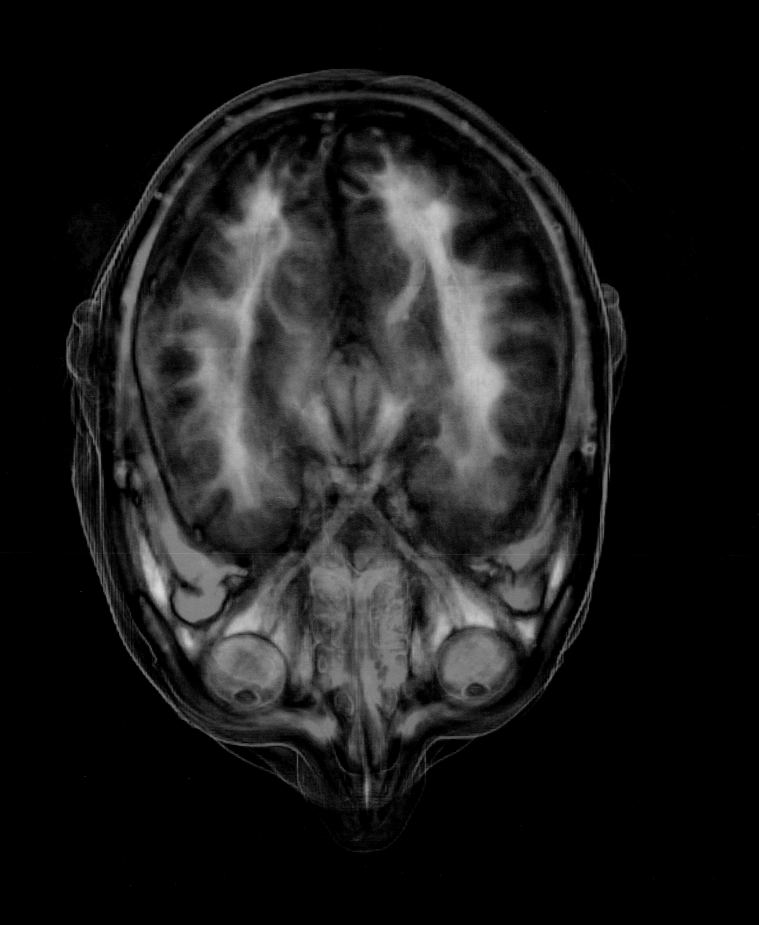

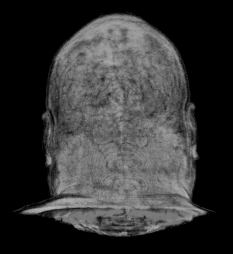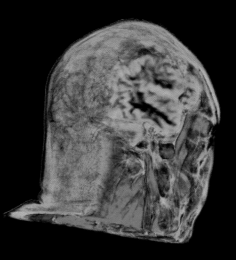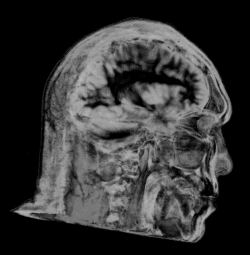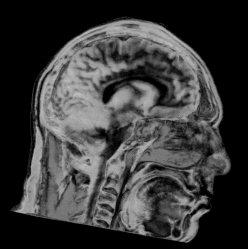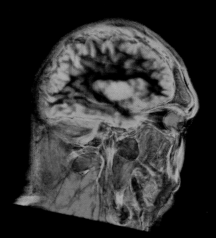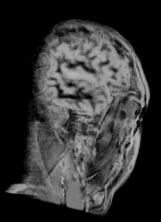

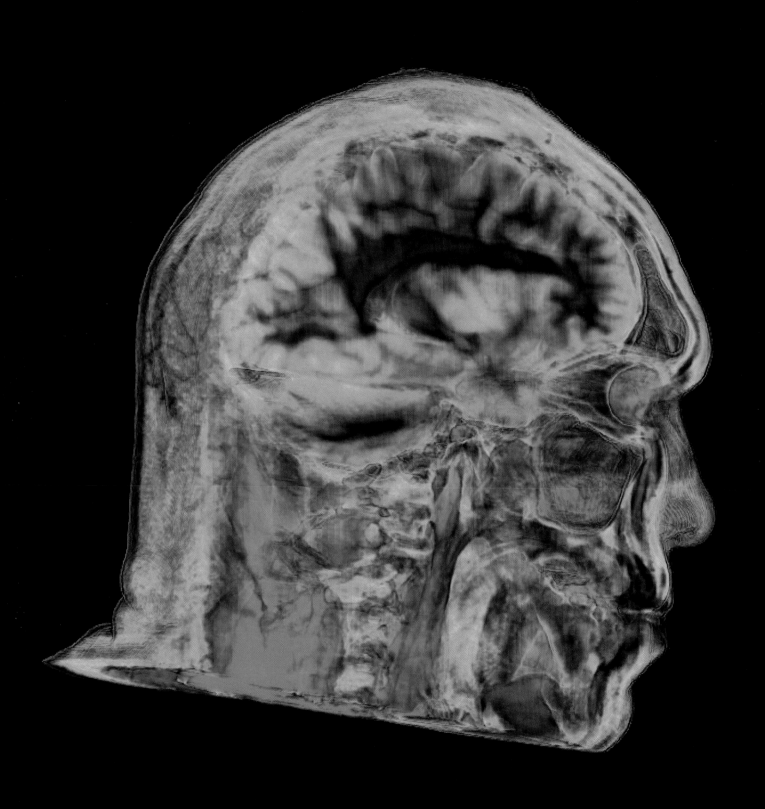

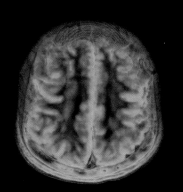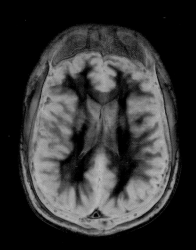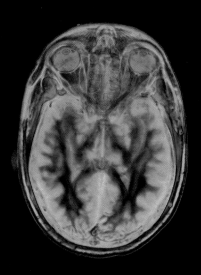
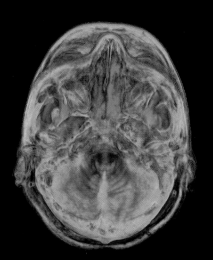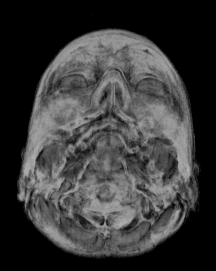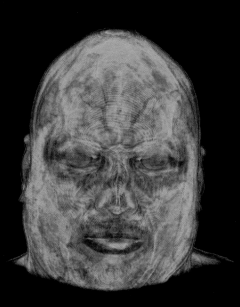

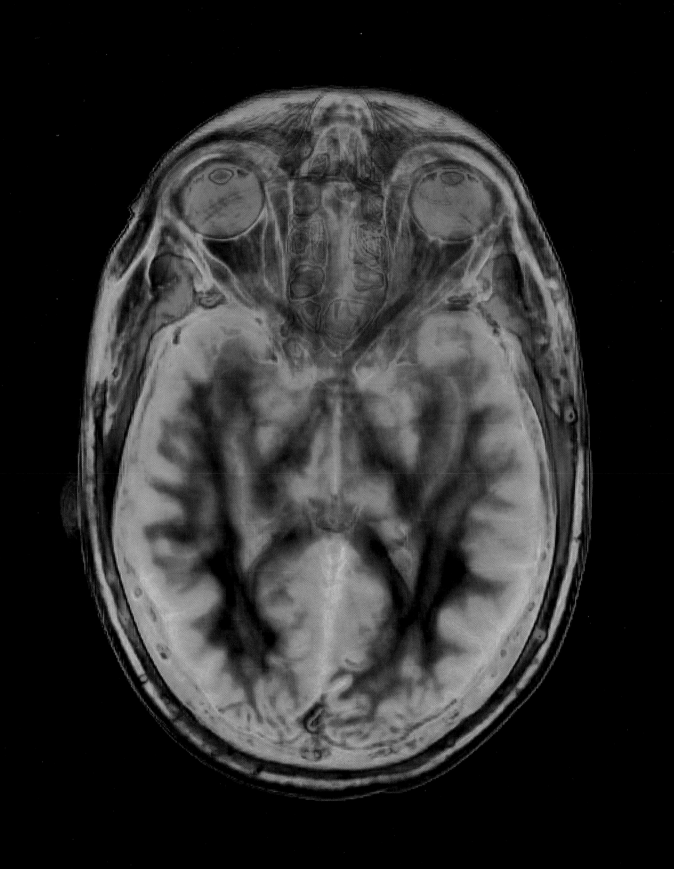

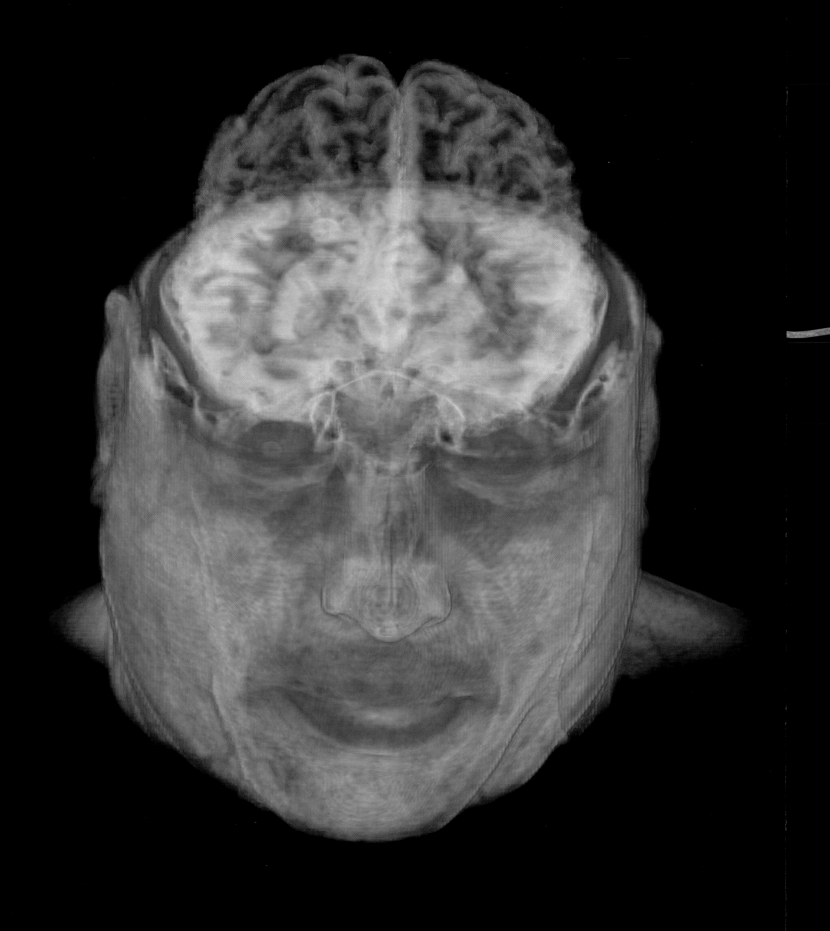

THE CHEMISTRY OF HEADACHES:

Author Christopher Buckley ranked the excruciating misery of his first cluster headache at 9 out of 10 on a scale of pain. By contrast, having a burning cigarette accidentally shoved into his eye rated only a score of 6. Such "splitting," "stabbing," or "goring" pain is a familiar fact of life for some 50 million Americans who seek medical treatment each year for acute or chronic headache. Yet, researchers are only beginning to understand the complex causes of severe headaches. The "cluster" headaches Buckley and over a million other Americans suffer are thought to be triggered when an upsurge of the chemicals serotonin and histamine in the brain causes hyperactivity and spasm in the trigeminal nerve, which extends from the forehead to the back of the neck. Migraine headaches, afflicting nearly 18 million Americans with episodes of throbbing pain, hypersensitivity to bright lights or loud noises, and nausea, have also been linked to serotonin surges—which can cause inflammation of arteries—in the large blood vessels at the base of the brain, the pial arteries in the back of the brain, and the parenchymal arteries that connect to them inside the brain. Serotonin, which causes the outer blood vessels to constrict tightly, cutting off oxygen to the nearby visual cortex, could also be responsible for the glowing "aura" many migraine patients report seeing prior to their attacks.

Physical or emotional stress can precipitate a serotonin surge in migraine and cluster headache sufferers. In the case of migraine, so can any number of other triggers, ranging from strong odors to fluorescent lights and the flicker of video display terminals. Tension headaches, the most common complaint of headache sufferers, also reflect the body's reaction to stress. When everyday anxieties and emotional conflicts get too intense, bands of muscles in the back of the neck, the forehead, and the temples contract. The result: viselike headache pain.

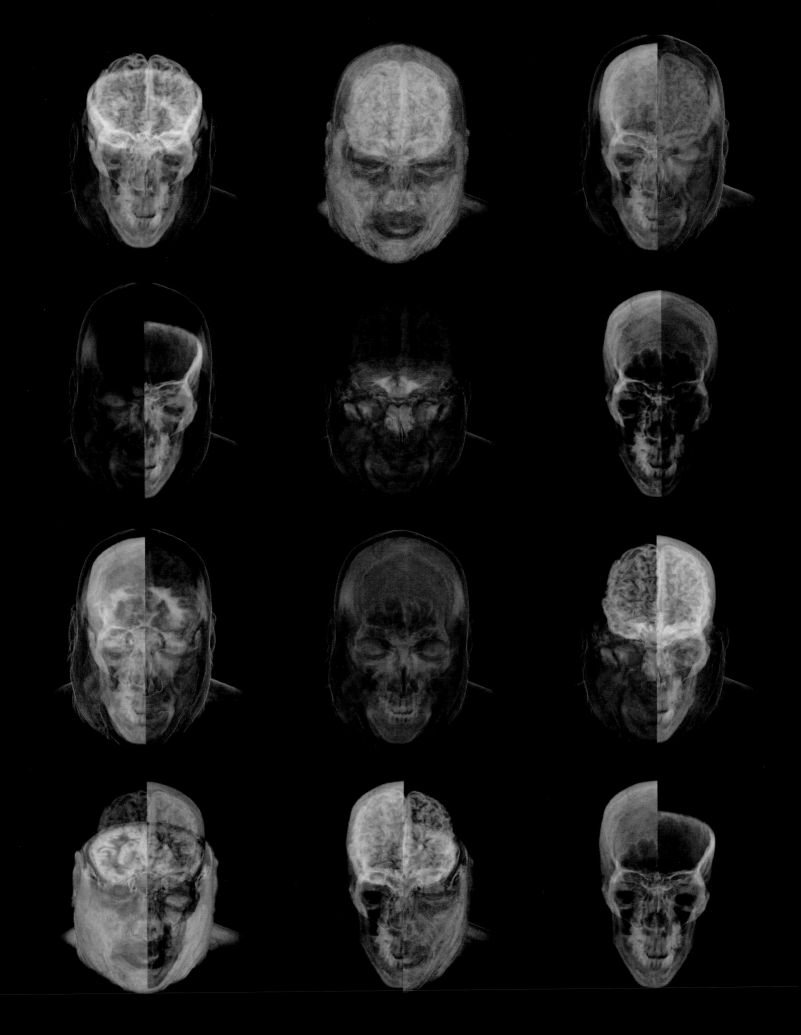

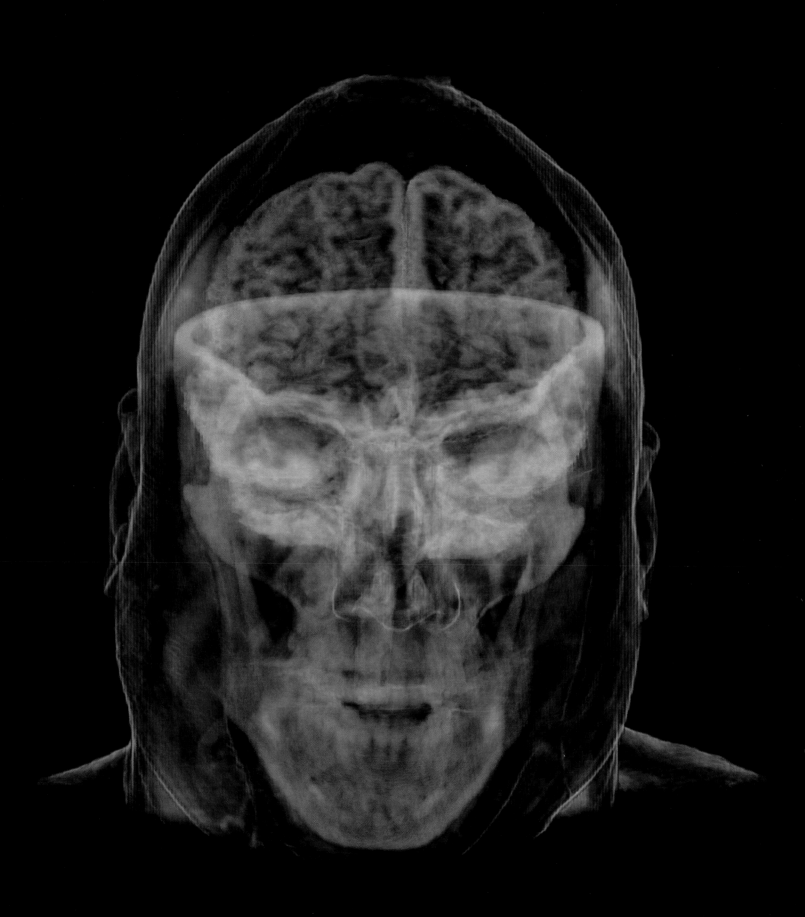

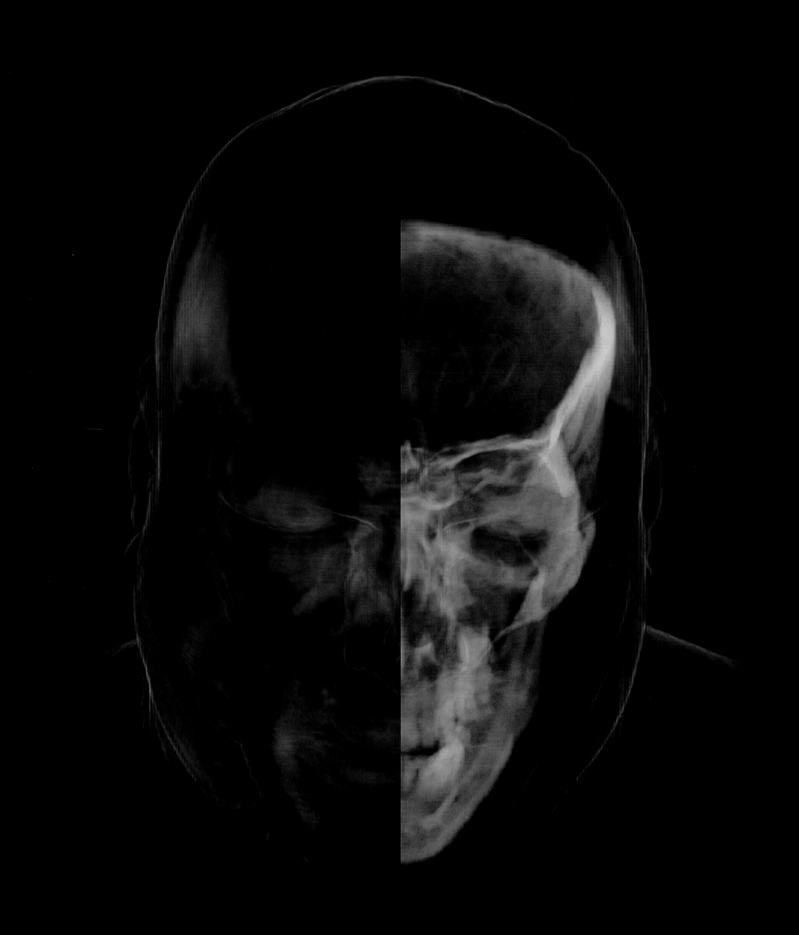

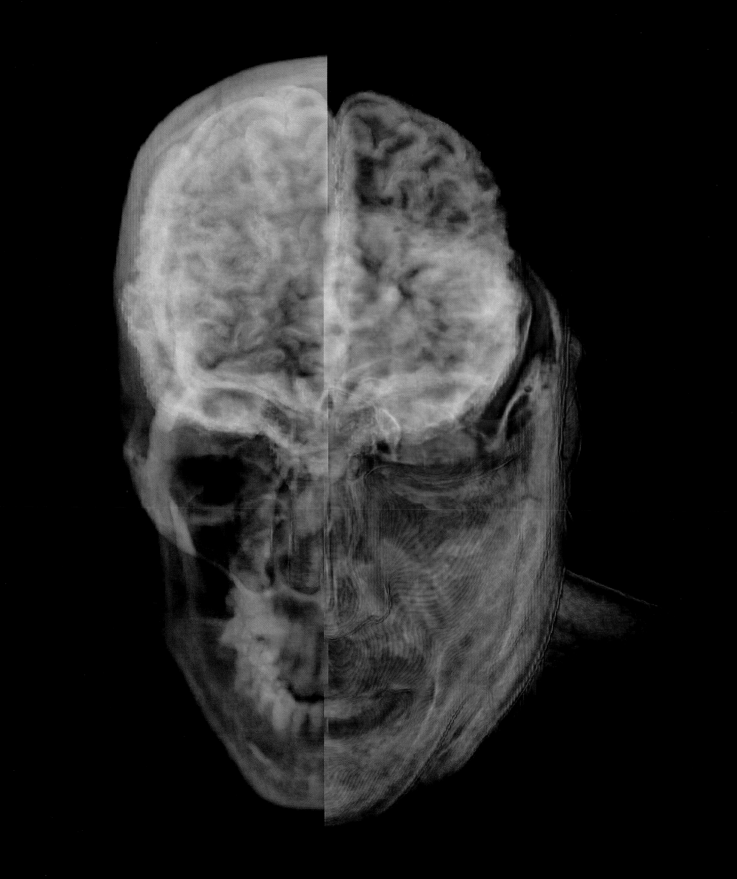

TWO BRAINS:

Psychologist Michael Gazzaniga once performed a revealing experiment on an epilepsy patient whose right and left brain hemispheres had been surgically severed to control his seizures. In the experiment, the Dartmouth University researcher flashed a picture of a chicken claw in the patient's right visual field (processed by his left hemisphere) and a picture of a snow scene in his left visual field (processed by his right hemisphere). He then tested the patient to see if he could identify pictures associated with the subjects he had just viewed. The patient responded correctly—choosing a picture of a chicken with his right hand and a picture of a snow shovel with his left. But when asked why he chose those items, the man replied, "Oh, that's easy. The chicken claw goes with the chicken, and you need a shovel to clean out the chicken shed."

To Gazzaniga, who has been studying such "split-brain" patients for over 30 years, the answer made perfect sense. In humans, the brain is divided into two hemispheres connected by the corpus callosum, a fibrous band of nerve cells that enables the two halves to communicate. Normally, the activities of the brain's two hemispheres are well integrated. In right-handed individuals, the left "dominant" hemisphere specializes in such rational pursuits as reading, language, writing, arithmetic, and other reasoning abilities, while the right hemisphere's functions are geared to more global, associative representations of experience that do not employ language—the ability to distinguish faces, for example. However, when the corpus callosum is cut, patients also lose the ability to transfer visual or cognitive information between the two sides of the brain. Thus, with its superior language and reasoning skills, the left hemisphere of Gazzaniga's split-brain patient was attempting to interpret the information in the pictures in a manner consistent with what it already knew—chicken feet, not snow. The patient could correctly perceive the images presented to each hemisphere, but his left brain was clearly doing the thinking.

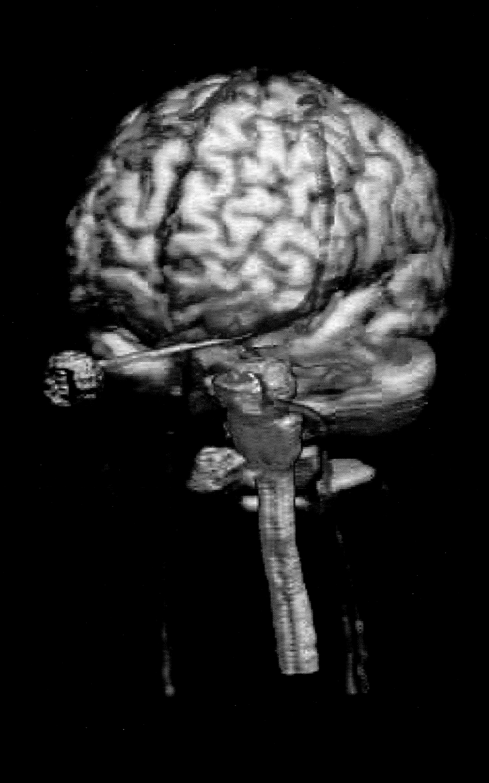

OPTIC NERVE

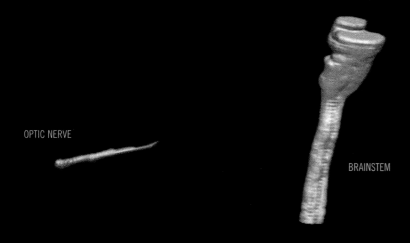

BRAINSTEM

HUMOR

EXTERNAL CAROTID

HUMOR

THALAMUS

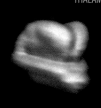

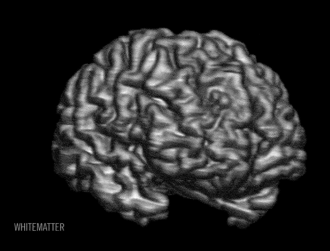

WHITEMATTER

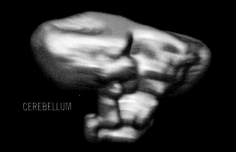

CEREBELLUM

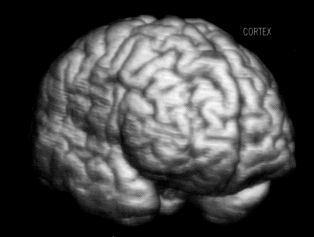

CORTEX

INTERNAL CAROTID

SUPER SAGITAL SINUS

OPTIC NERVE

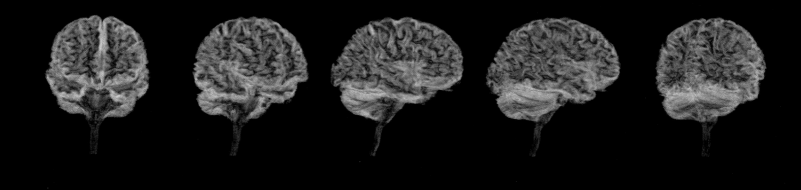

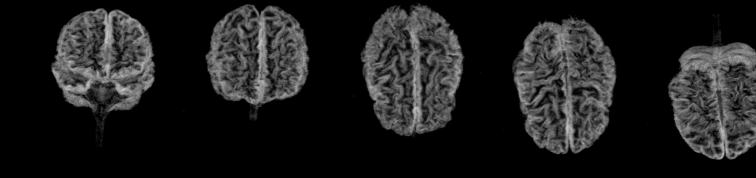

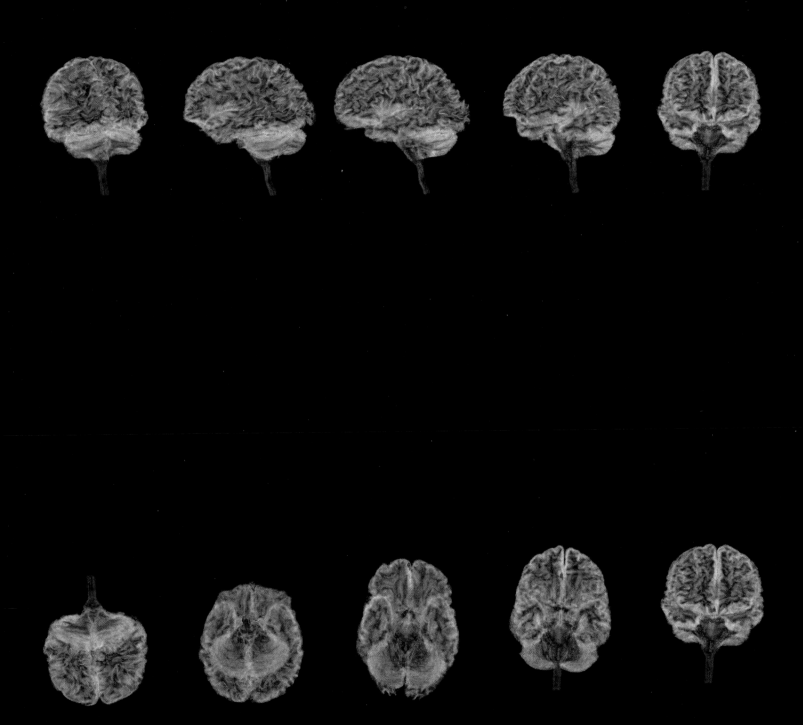

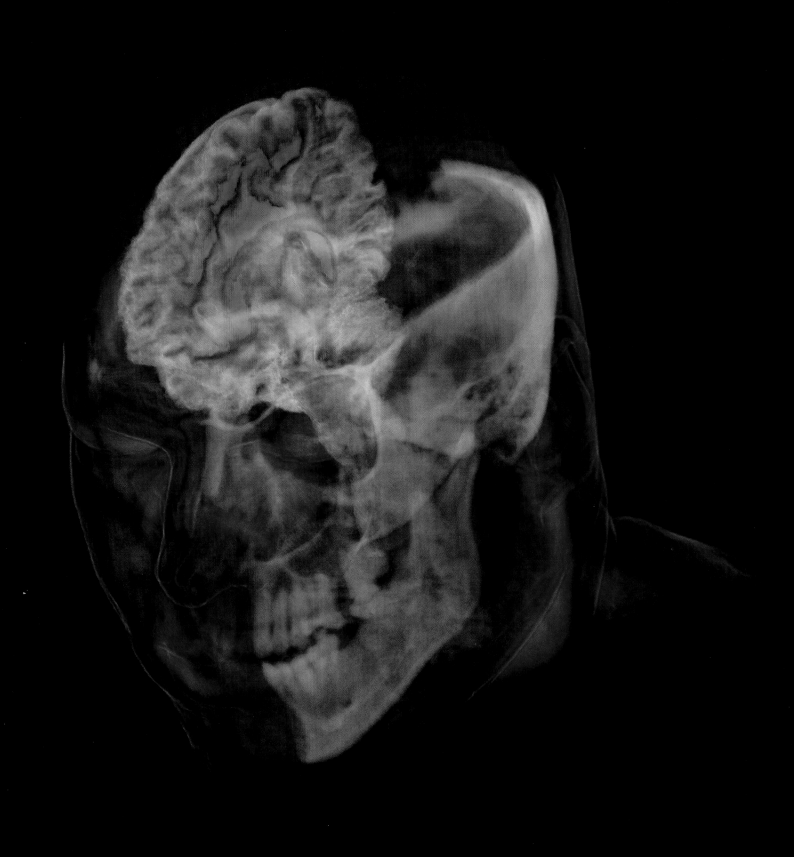

THE ARCHITECTURE OF MEMORY:

The life of the middle-aged newspaper advertising salesman known as "Boswell" was limited to a handful of facts: his name, his occupation, his military service, and the faces and names of his close family members. The rest of his vast autobiographical web of memory had been swept away in the wake of a profound 3-day coma caused by encephalitis, an inflammation of the brain produced by the herpes simplex virus. Nor was he able to form a single new memory in the 8 years that followed. Events occurring even 5 minutes earlier were promptly enveloped in the ever-recurring, ever-fading moment in which Boswell existed.

Much of what we know about where the brain processes memory comes from studying amnesiacs. While the loss of memory can also be symptomatic of a more widespread deterioration of brain cells, such as is found in the late stages of Alzheimer's disease, the pure forms of amnesia exhibited by patients like Boswell are usually caused by injuries to specific brain regions. For example, the "aphasias"—puzzling losses of the ability to speak or understand speech—have long been associated with injuries to regions in the lower frontal lobe and the posterior temporal lobe of the brain's left hemisphere, which are thought to be storage centers for language memory. More recently, researchers studying amnesic patients have identified damage to the hippocampus—one of the parts of Boswell's brain that was destroyed—with an inability to learn or store new memories. Lesions of the forebrain, Boswell's other major site of injury, often produce "mismatched" recall, an inability to connect names and faces. Injury to the amygdala and certain sections of the right hemisphere can disrupt the ability of patients to recognize faces, including their own.

Even our sense of morals may be a kind of "social" memory, with its own repository in the brain. One patient, described by the neurologist Antonio Demasio in his book *Descartes Error,* appeared to lose his memory of social norms when much of his frontal lobe was ablated to remove a tumor. After his recovery, the previously frugal, church-going accountant was transformed into a philandering, argumentative wastrel.

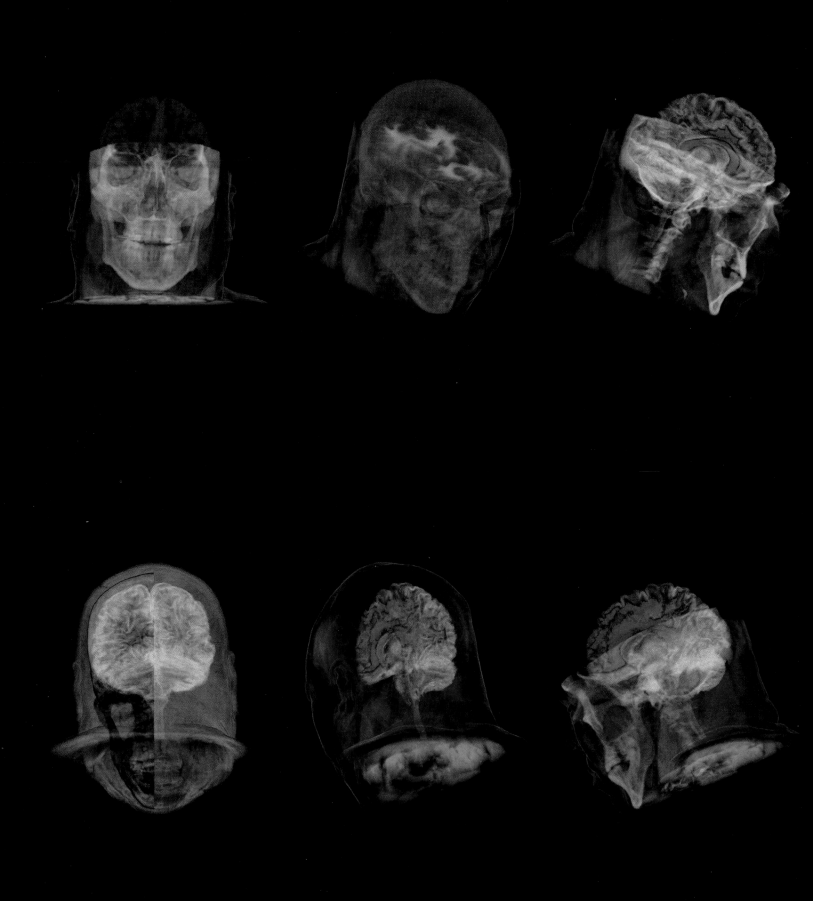

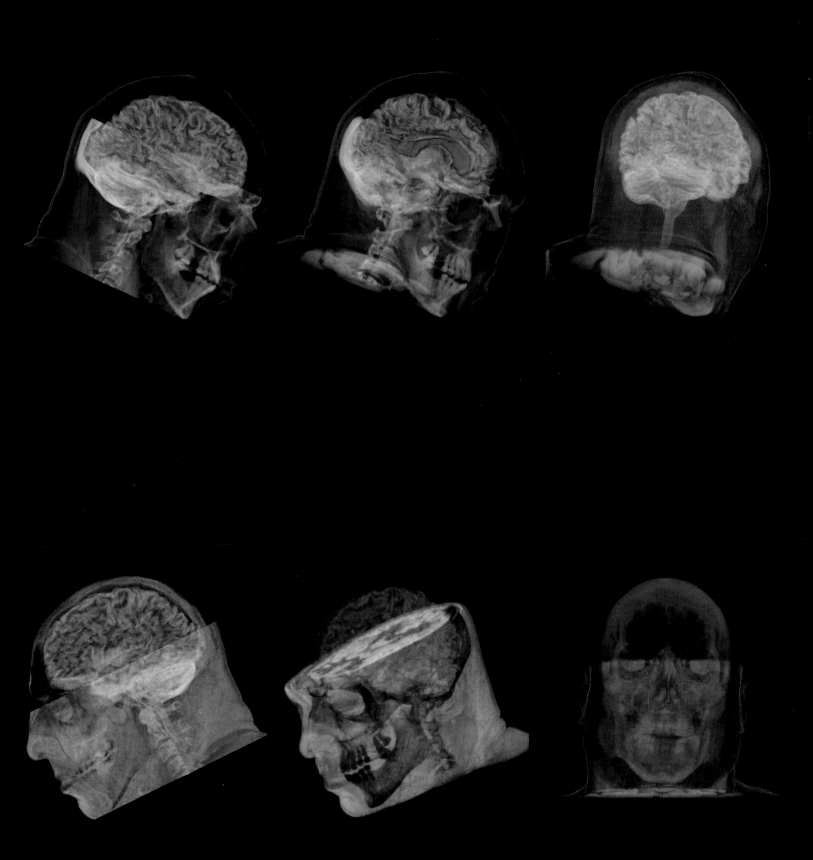

HOW THE NOSE KNOWS:

"We think we smell with our noses [but] this is a little like saying that we hear with our ear lobes," wrote Gordon Shepherd, a neuroscientist at Yale University.

In fact, the part of the nose we can see from the outside serves only to take in and channel the air containing the chemicals of fragrance called odorants. The neurons that actually sense these molecules lie at the very top of the nasal cavity in a thumbnail-size patch of cells called the olfactory epithelium.

Each olfactory neuron in the olfactory epithelium has a long fiber, or axon, that pokes through a tiny opening in the bony cribriform plate above it to connect to other neurons in the olfactory bulb—a round, knoblike brain structure, which is largest in animals, such as bloodhounds, that have a sharp sense of smell. From there, the olfactory tract projects to regions of the limbic system and frontal cortex, including the hypothalamus, a key structure that controls sexual and maternal behavior. This anatomical arrangement might account for the sensuous allure of perfume or the fact that even human mothers can recognize their babies by smell, but it does not explain how the average human being can recognize some 10,000 separate odors.

The answer may lie, in part, in the sheer diversity of olfactory nerve cells. Studies analyzing human DNA have identified over 100 different genes producing individual types of olfactory neuron—each one thought to be sensitive to one or a small combination of different odorants. Scientists suspect that thousands more distinctive odor receptors may contribute to our sense of smell.

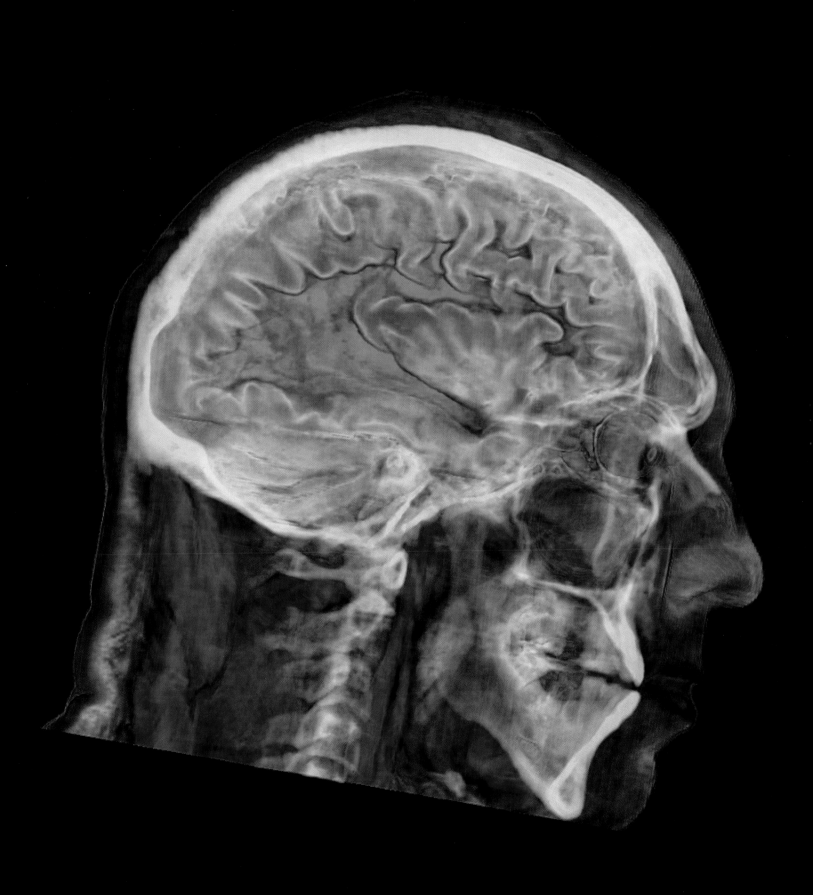

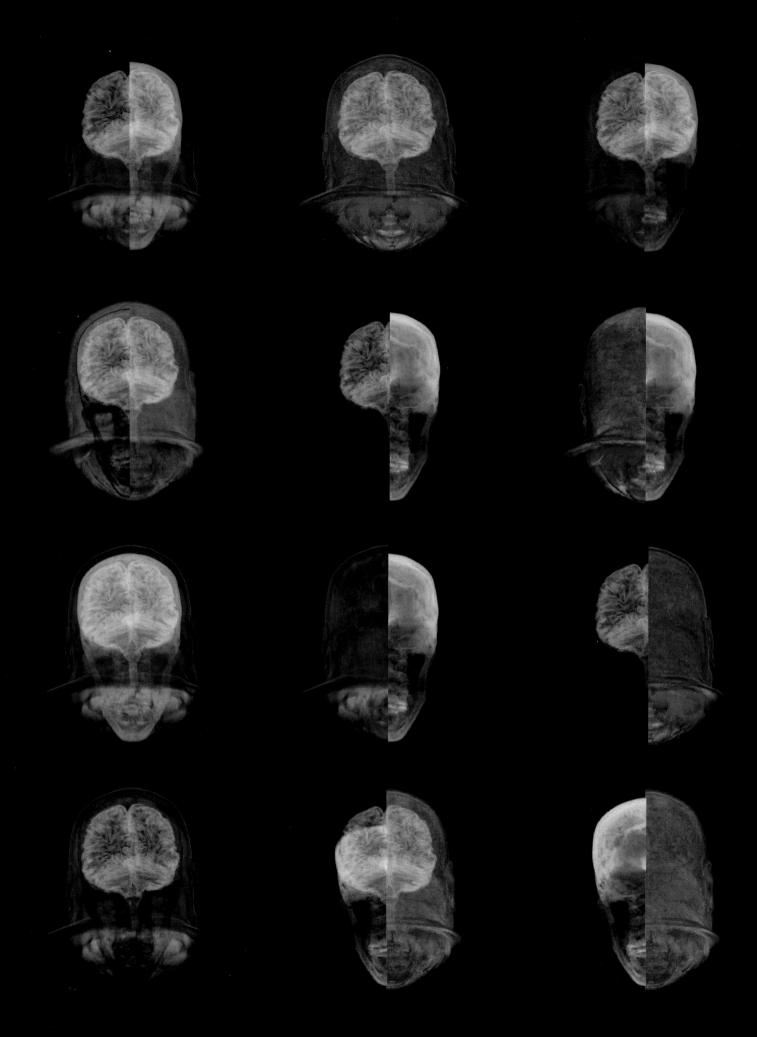

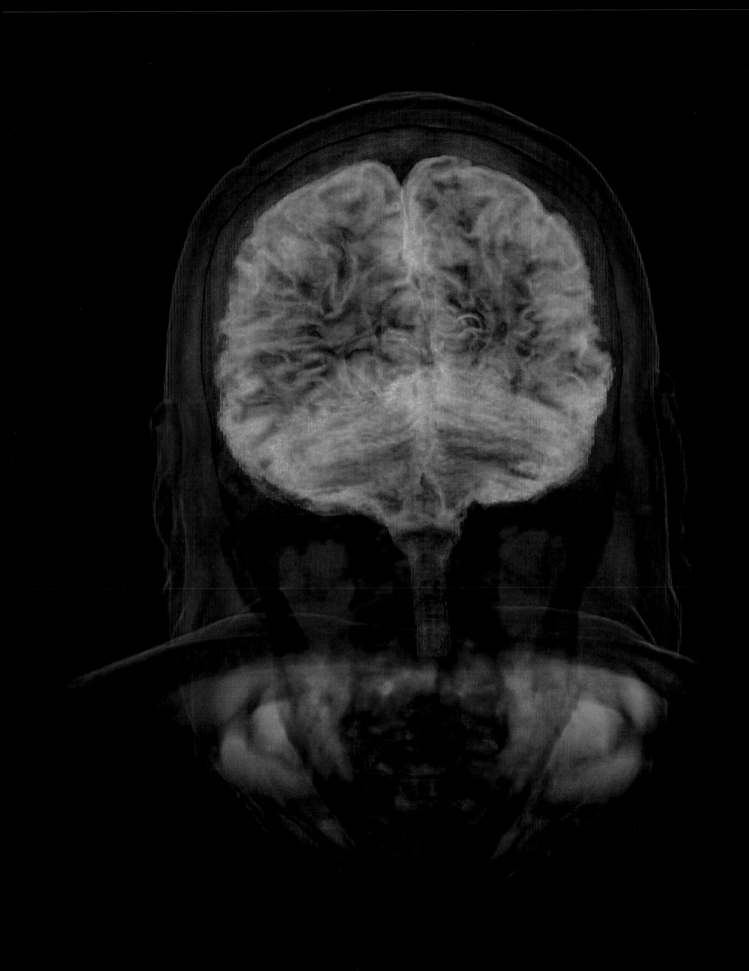

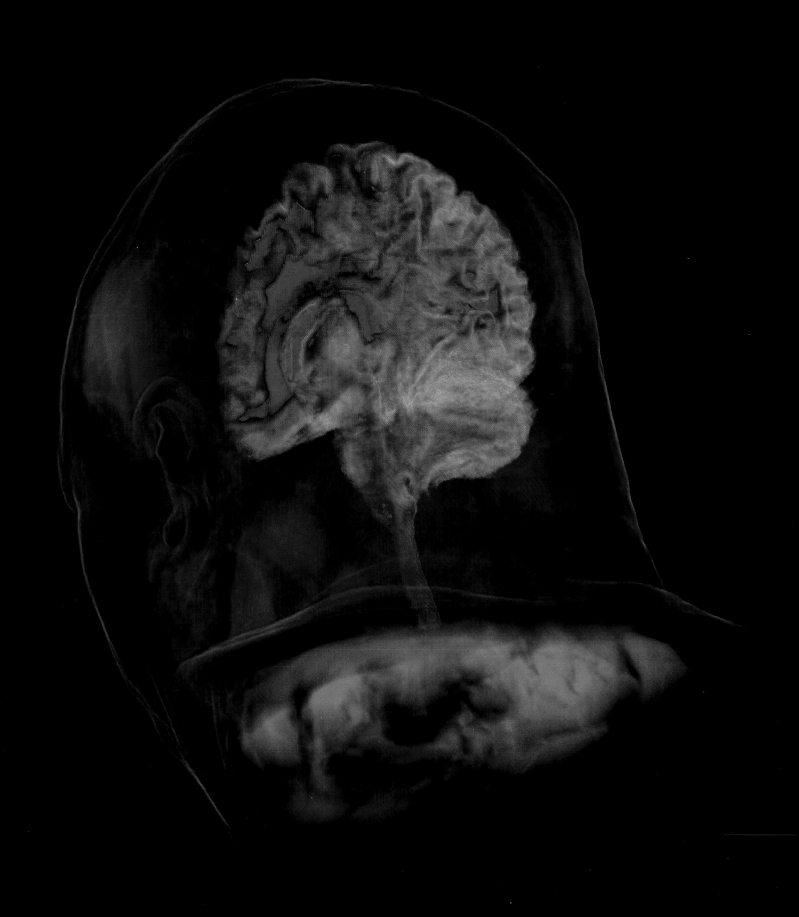

A KEY COMPONENT OF SPEECH:

A larynx or voice box positioned low in the neck is a distinctively human trait that enables us to create a wide variety of sounds. In other mammals, the larynx is positioned high in the neck and locks into the nasopharynx—the air space at the back of the nasal cavity—during breathing. This arrangement has some advantages, allowing nonspeaking mammals to drink and breathe at the same time, but it limits the array of sounds an animal can produce.

According to fossil evidence, the evolution of a lower larynx may have facilitated the birth of language nearly 1 million years ago, at the time of the earliest known human ancestor, Homo erectus. Interestingly, this same evolutionary pathway, from nonspeaking to speaking animals, is re-created as every child grows. At birth, an infant's larynx locks into the back of the nasal cavity, forming a kind of protective barrier between the breathing and swallowing pathways. After 18 months, an infant's larynx begins to descend in the neck, and the pharynx expands. The transformation enables the child to create a broader range of sounds but leaves a large, exposed space above the larynx, where food and airways now cross. The anatomical arrangement gives new meaning to the admonition, "Don't eat with your mouth full." Though a lower larynx may enable us to speak, food can lodge in the throat, blocking the airway and causing choking.

"Don't Speak with your mouth full"

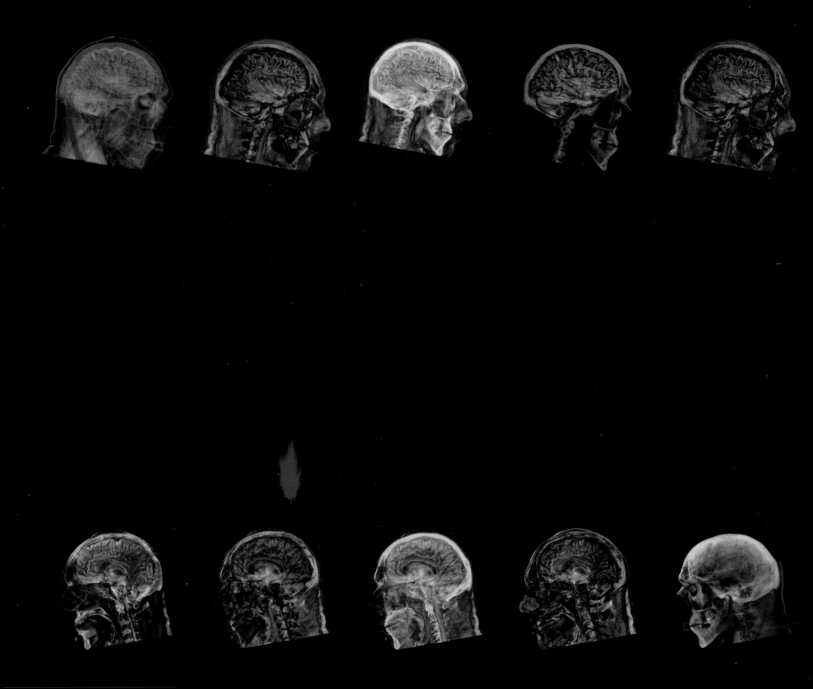

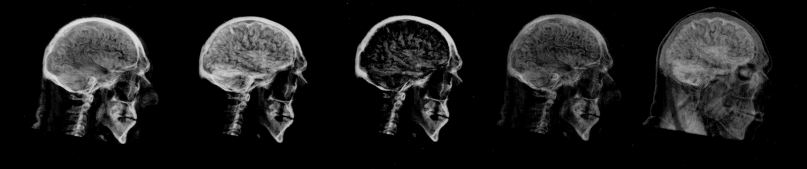

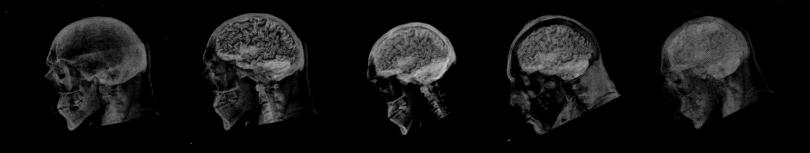

OH, MY ACHING JAW:

Some 10 million to 20 million people have problems involving jaw pain, accompanied by a grinding or clicking of the jaw when they talk or eat. Jaw muscles become sore, chewing is difficult, and pain spreads to the facial and neck muscles and persists around the clock. Such syndromes, called temporomandibular disorders (TMD), occur mainly in women and are often caused by teeth clenching and grinding (bruxism), associated with emotional distress. Bad posture, whiplash injuries, and chewing too much gum can also promote or aggravate TMD. While over-the-counter pain medications, stretching exercises to relax sore muscles, or a change in diet will usually alleviate the problem, persistent temporomandibular disorders have driven thousands of sufferers to seek surgical solutions for their pain—often with disastrous results. In the late 1980s, between 75,000 and 100,000 TMD patients underwent operations in which Teflon-laminated implants, about the size of a fingernail, were inserted into the joint at the point where the jaw bone joins the skull. The implants, which were supposed to act as a kind of shock absorber, not only failed to alleviate symptoms but damaged bone and left some patients permanently disfigured and disabled.

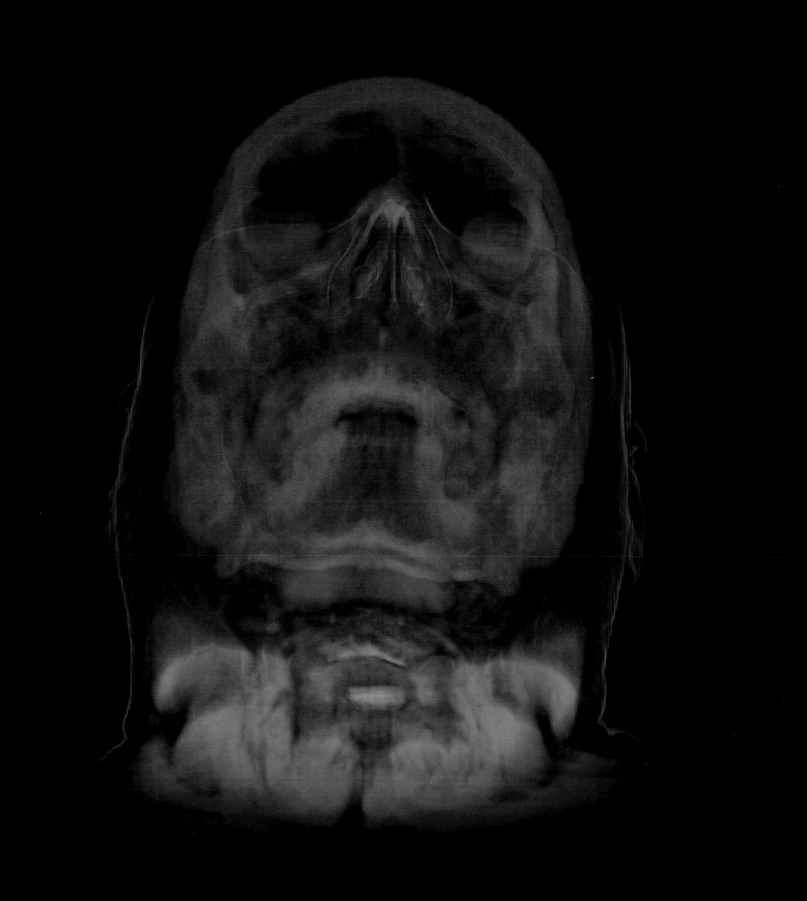

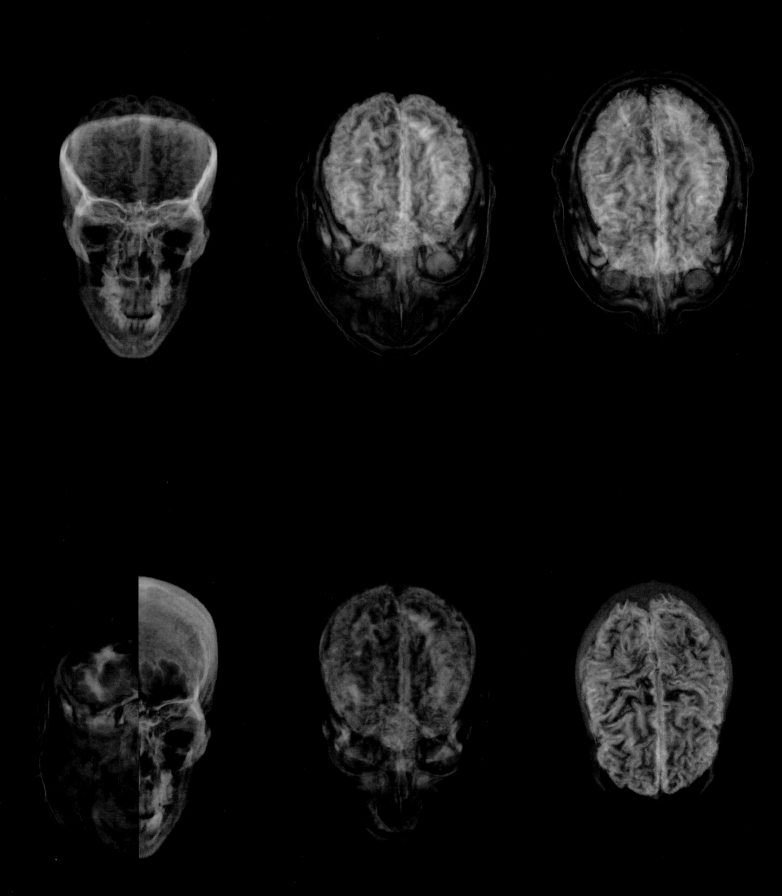

ANATOMY OF THE COMMON COLD:

Scientists may not be any closer to a cure for the common cold, but a 1994 study by investigators at the University of Virginia Health Sciences Center has revealed that cold viruses stage a more wide-ranging attack and are more tenacious than previously suspected. The researchers used computed tomography (CT) scans, normally reserved for detecting head injuries or suspected tumors, to peer inside the heads and upper respiratory tracts of 31 cold sufferers. As expected, the examinations typically revealed thickened nasal passages (ethmoidal infundibulum) and inflammation of the pharynx. But in a surprisingly high proportion—87 percent—colds had also spread to one or both of the maxillary sinus cavities, which drain into the nose. A stuffy nose and sore throat are typical cold symptoms, but physicians have long regarded any extension of bacterial infections to the paranasal sinuses as a complication. Based on the new findings, the Virginia researchers have reclassified the common cold as a viral rhinosinusitis, capable of infecting the sinuses, as well as the nasal passages and the throat. The researchers also observed that, while cold sufferers may feel much better after a couple of days in bed, it takes considerably longer for the immune system to combat the virus. Follow-up CT scans revealed that nearly half still exhibited upper respiratory abnormalities as much as 2 weeks after their initial examination.

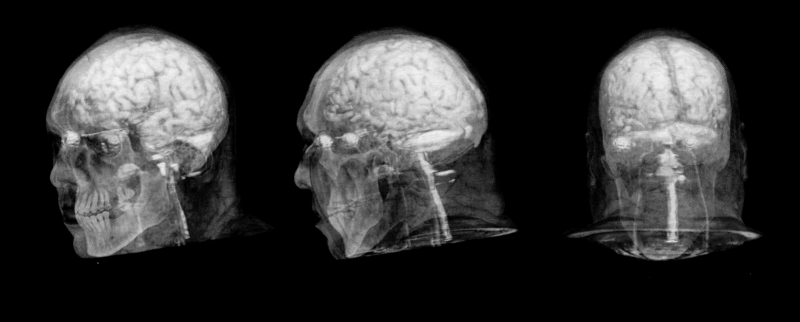
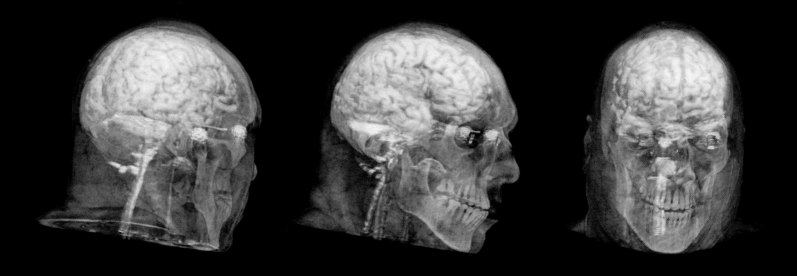

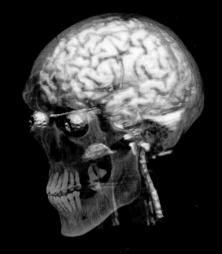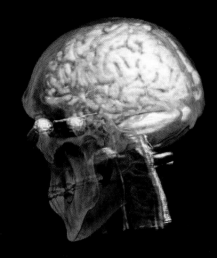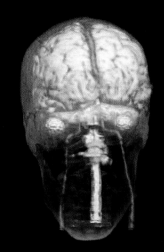

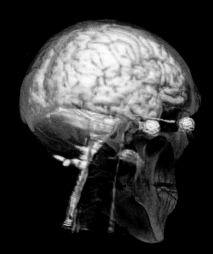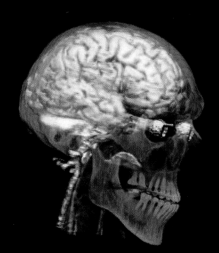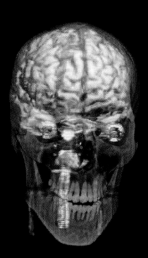

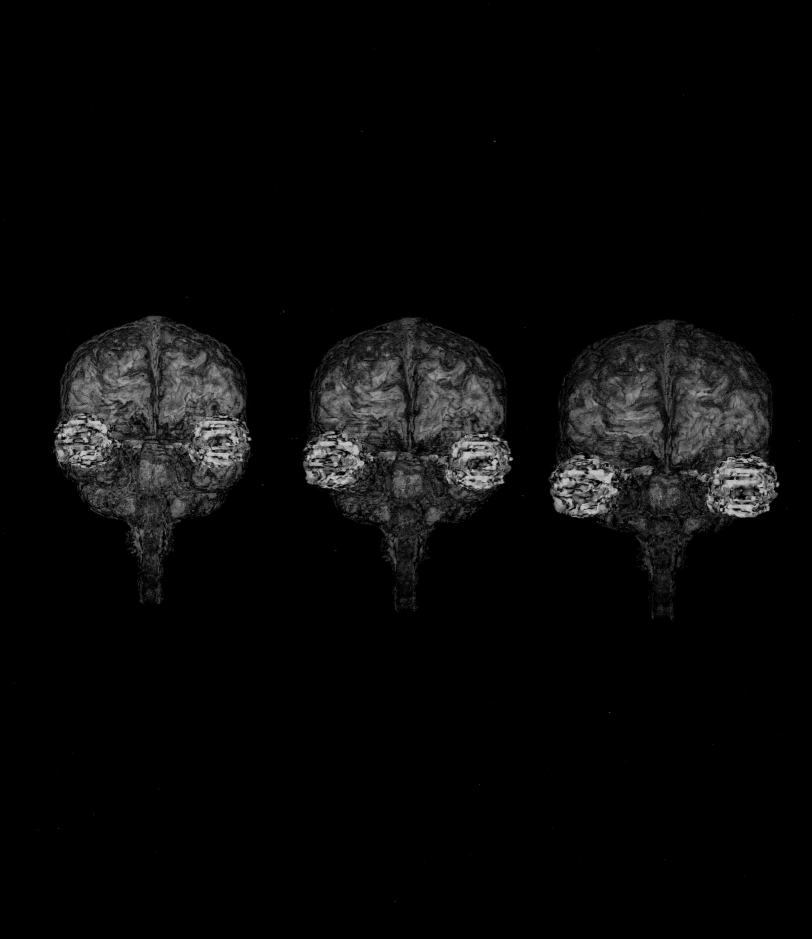

MAPS OF VISION:

The world was frozen into a series of stop-motion photographs for the patient at the Max Planck Institute of Psychiatry in Munich, Germany. A stream of water pouring from a faucet looked like a chunk of ice; a car approaching from a distance was suddenly upon her. A magnetic resonance imaging (MRI) scan revealed that the woman's motion blindness resulted from a stroke, which had destroyed portions of her midbrain. She was otherwise unimpaired and could see normally, without glasses.

The woman's inability to perceive movement resulted because the visual pathway, which involves about a quarter of the human cerebral cortex, is amazingly specialized. The primary visual cortex or area V1, which curves around a deep fissure at the back of the brain, possesses a highly organized system of neurons for analyzing the orientation of an object's vertical and horizontal outlines. In an adjacent area called V4, cells burst into activity at the site of different colors. By probing the visual centers of laboratory animals with fine electrodes to pick up firing in response to movement, researchers have found that parts of the visual system are also equipped with motion detectors—specialized cells that respond only to movement in a particular direction, not to the color or form of an object. These sections, called MT (middle temporal area) or V5, are located just beyond the primary visual cortex—in precisely the same areas that were damaged in the German patient suffering from motion blindness.

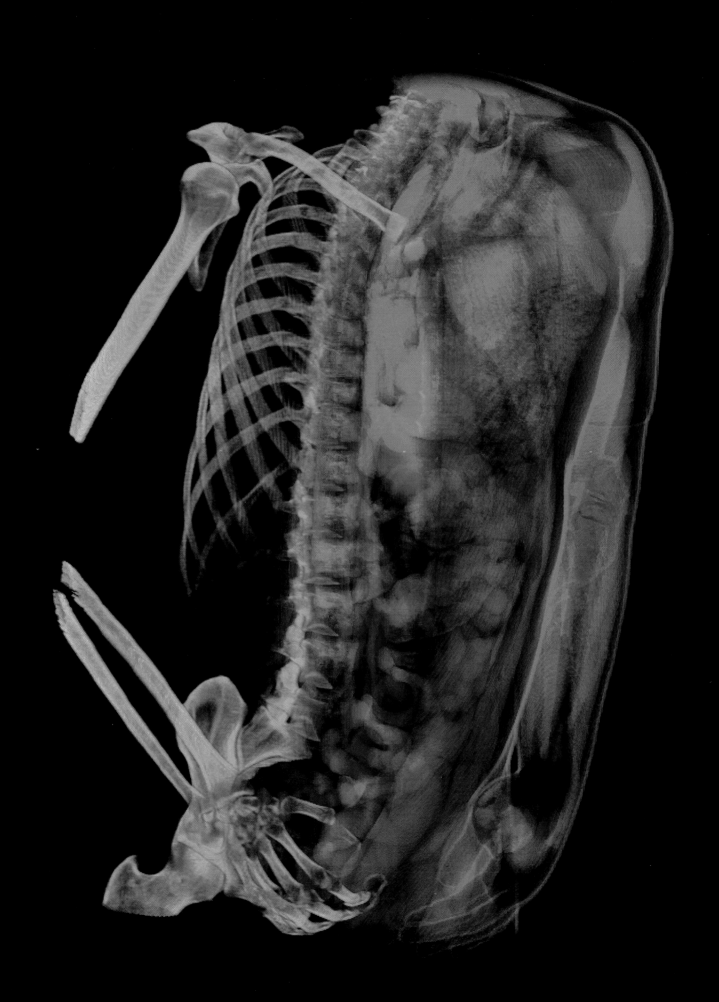

The Torso

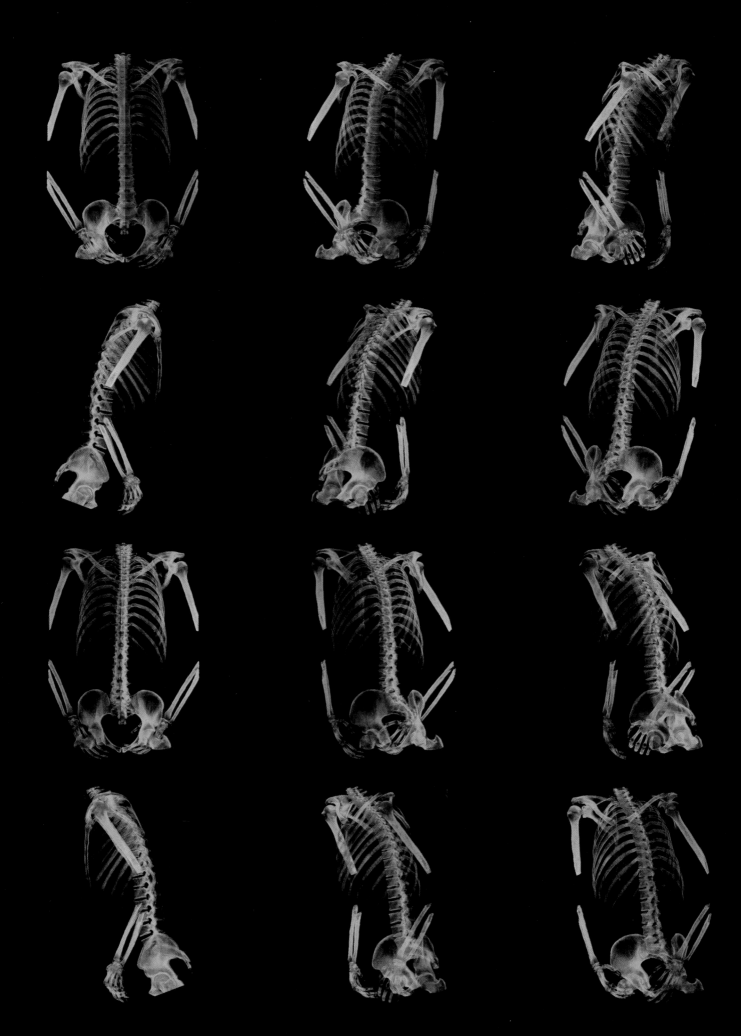

If, as Emerson wrote, "the human body is the magazine of inventions," nowhere is this observation more apparent than in the torso, where the wondrous engines, pumps, bellows, and plumbing of the body are housed.

Even before the precise functions of these organs were fully understood, they were invested with extraordinary powers. The fluttering heart—capable of pumping the equivalent of 2,000 gallons of blood a day through the body's 600,000 miles of blood vessels—was regarded by ancient philosophers as the seat of the soul and the engine of sentiment and love. The liver—now known to be an amazingly sophisticated chemical warehouse and factory—was regarded as the font of combative or melancholic humors. Victorian physicians extolled the state of the digestive organs as the oracle of health and the harbinger of disease.

Today, doctors rely on centuries of experience to diagnose and treat these life-sustaining machines when they fail or break down—whether they are repairing a failing heart with bypass surgery or deciding whether a bellyache is due to an upset stomach, appendicitis, or one of dozens of other possible causes. With keen observation, tireless experimentation, and sometimes sheer luck, breakthroughs—like the recent discovery that a treatable bacteria causes most stomach ulcers—and a steady increase in knowledge, have enabled physicians and surgeons to reach hitherto unimaginable levels of achievement. Livers and kidneys are now routinely transplanted with high rates of success; researchers are well on their way to perfecting "artificial" blood and a portable, battery-powered artificial heart; and "gene" therapy techniques are being developed that may soon enable physicians to treat deadly inherited diseases, such as cystic fibrosis, by implanting healthy DNA into abnormal cells.

The price of progress: a host of perplexing ethical issues

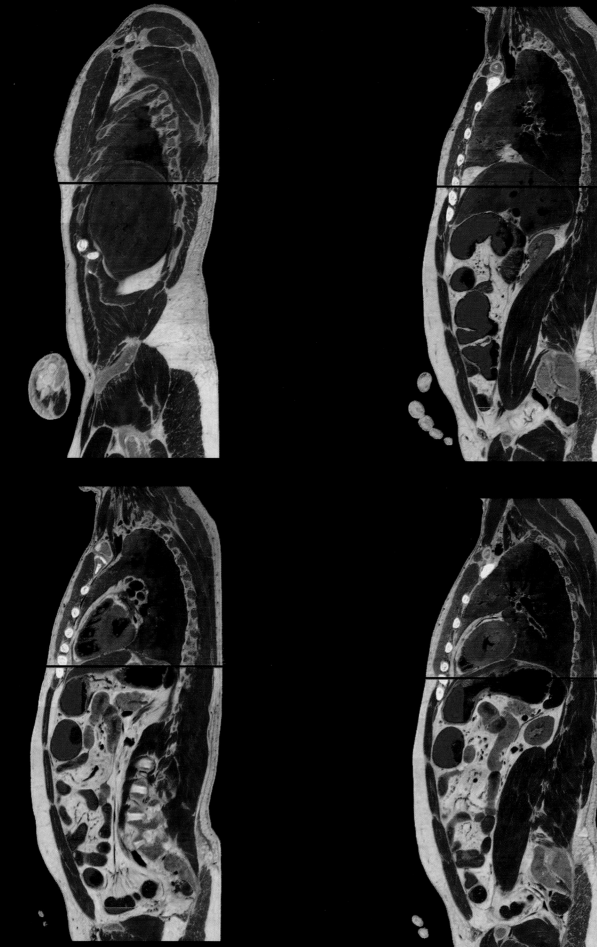

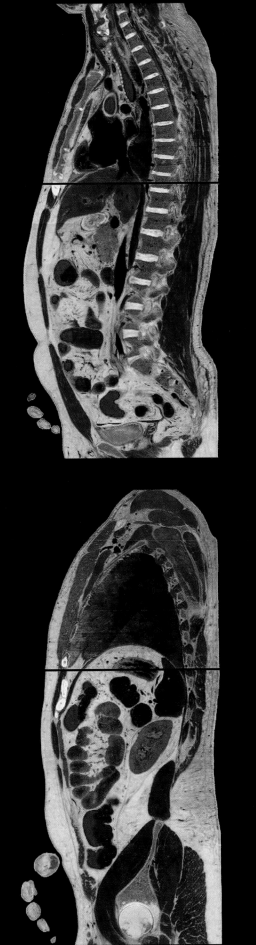
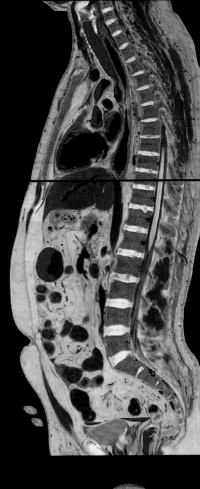
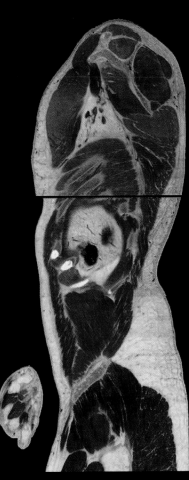

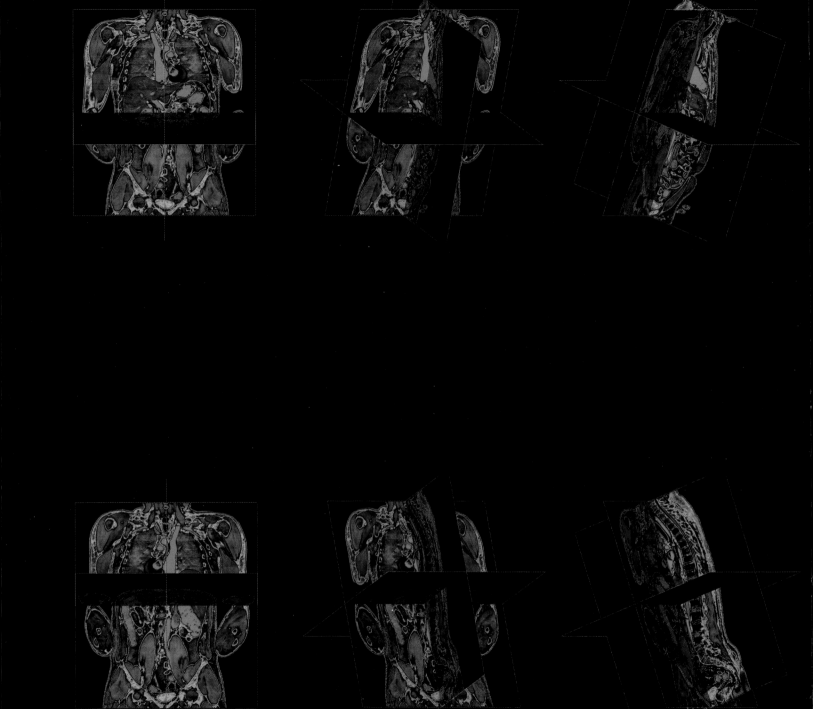

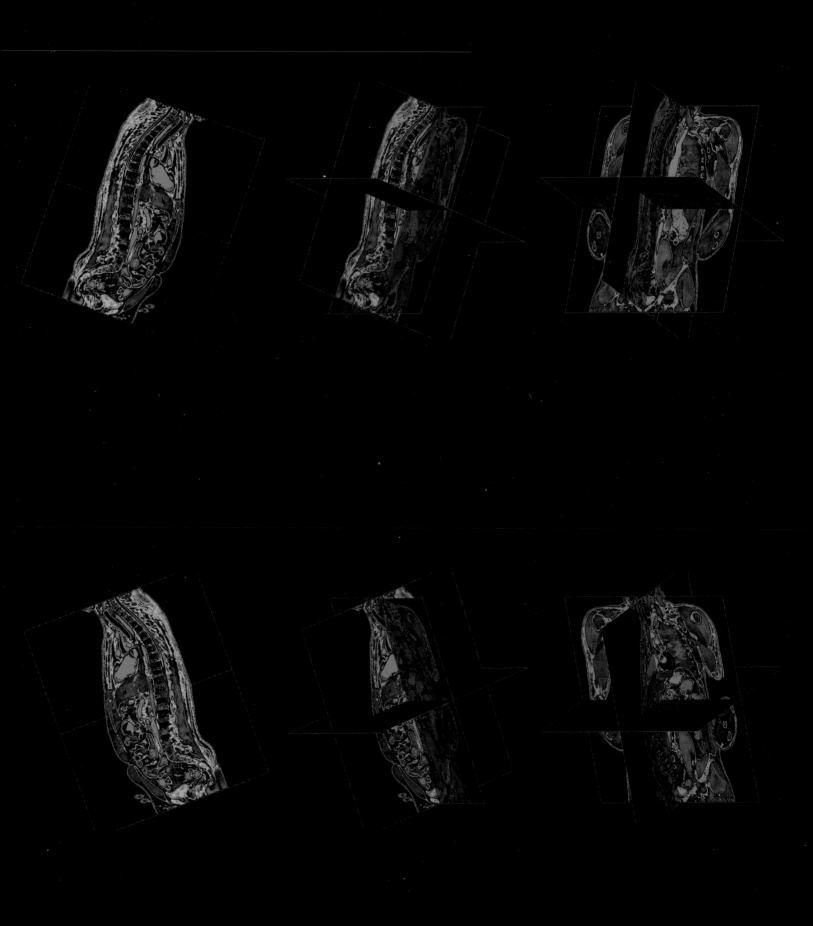

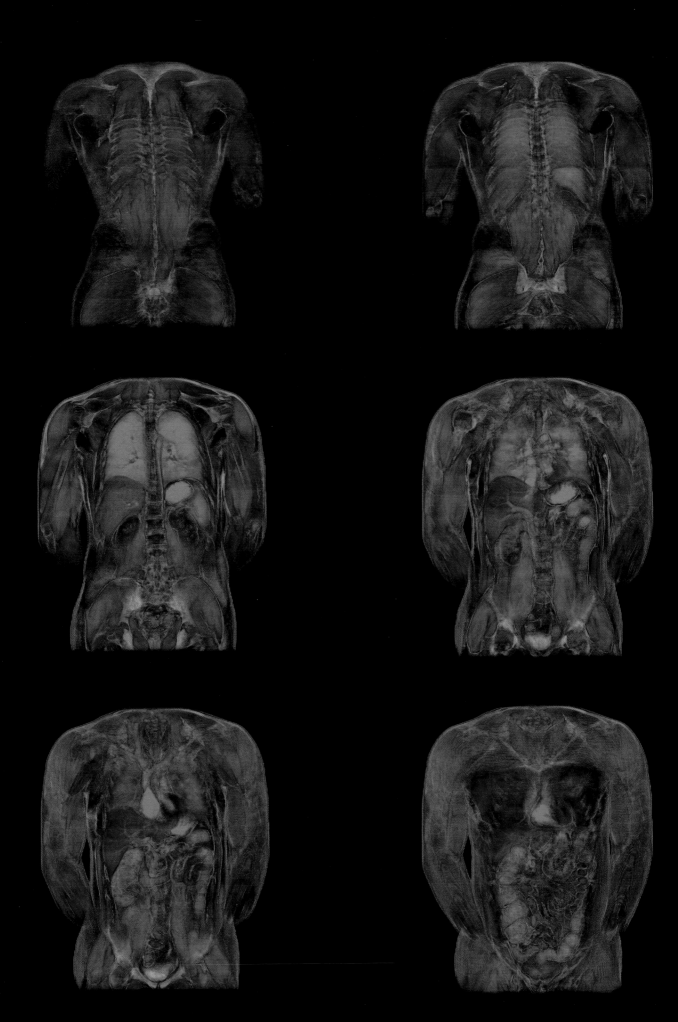

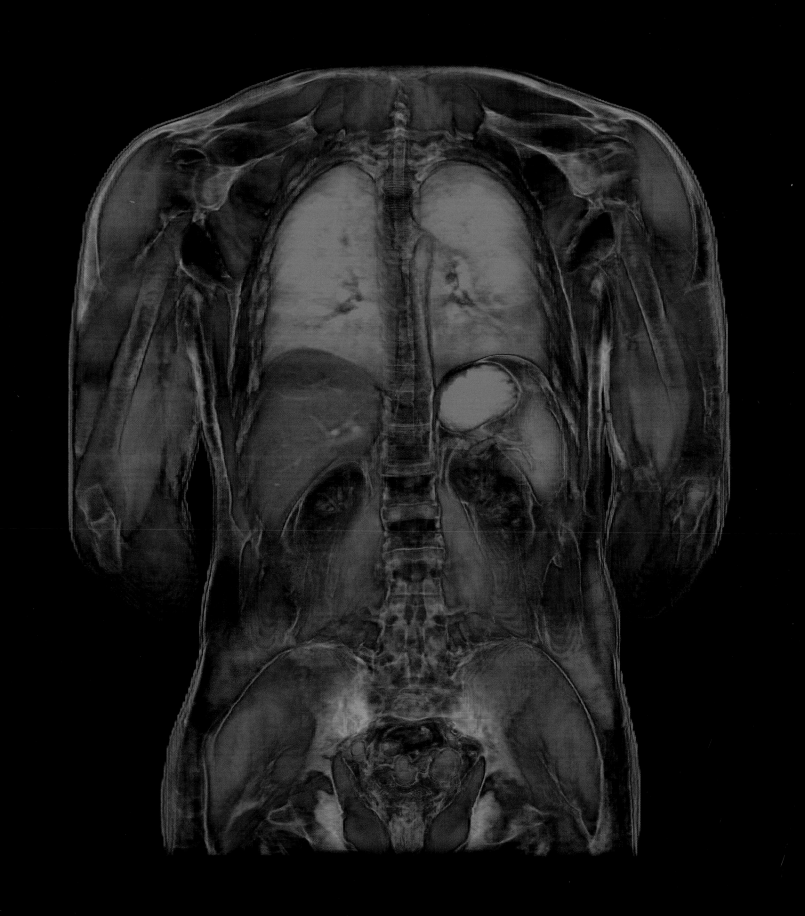

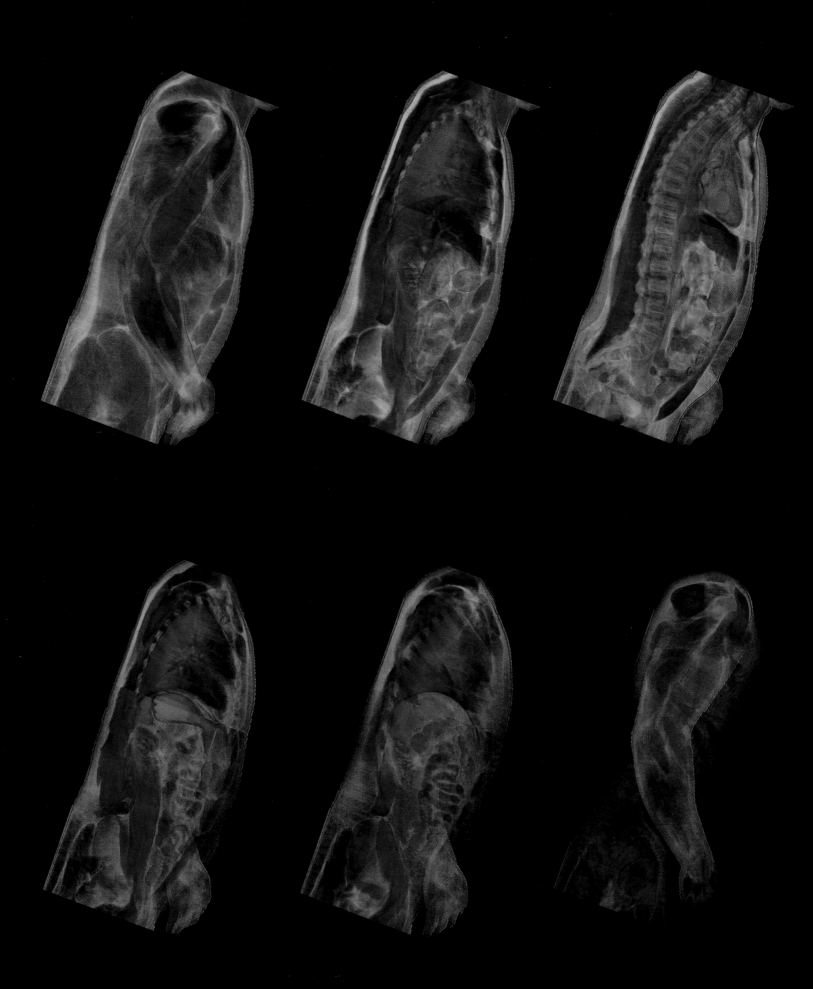

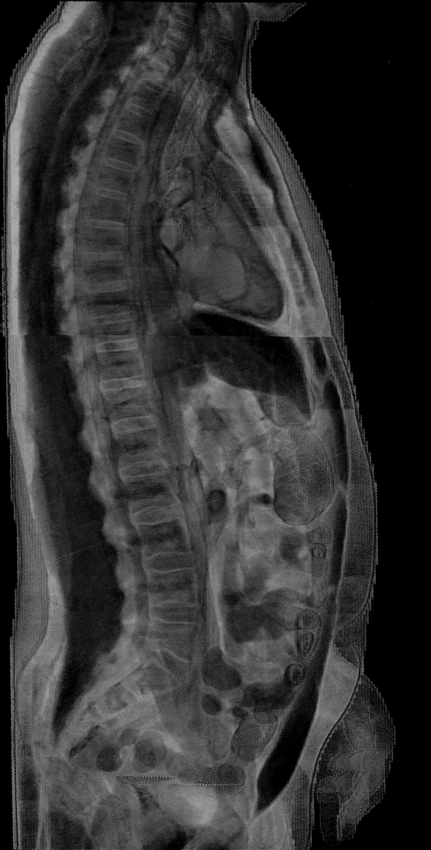

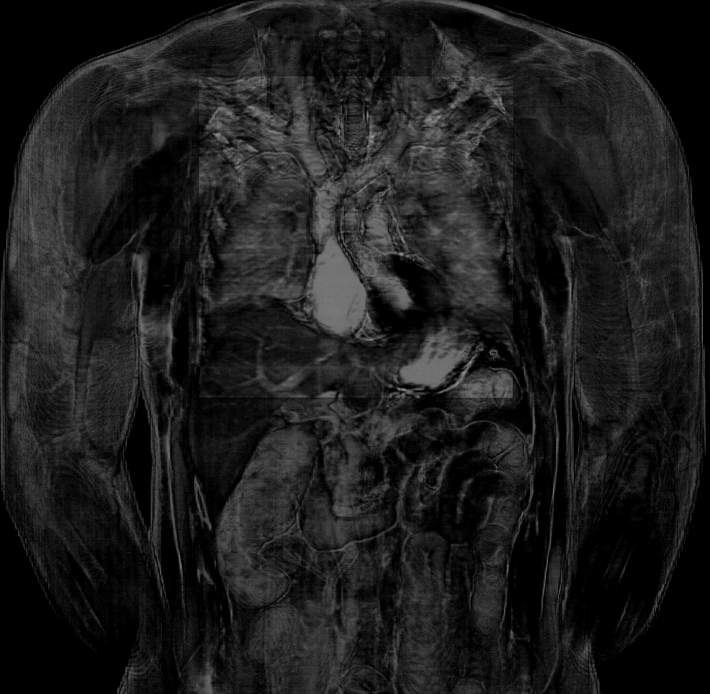

ARTIFICIAL BLOOD:

The patient was rushed to the emergency room of Denver General Hospital after a car crash, bleeding to death and in desperate need of a transfusion. But when tests revealed that he had an extremely rare blood type, which could take crucial hours—or even days—to match, his doctors were ready with an alternative. Instead of waiting, a transfusion of 4 units of a laboratory-produced blood substitute, called PolyHeme, was prepared. The patient stabilized and survived.

Since the seventeenth century, scientists seeking blood substitutes have tested everything from milk and vegetable oil to beer. But today, for the first time, researchers are close to perfecting an oxygen-carrying (and completely safe) "artificial" blood to treat critical blood loss from accidents or surgery. Half a dozen blood substitutes are currently undergoing human trials. The man-made products—composed of the oxygen-carrying component of blood, hemoglobin, synthesized into strings called polymers—have several potential advantages over natural blood. They do not require refrigeration and have a shelf life of up to a year, compared to 28 to 42 days for donated blood; they can be sterilized, rendering them virtually free of contamination from viruses, such as HIV or hepatitis; and because they require no blood typing, they can be used as a short-term substitute for the estimated 10 percent of transfusion patients, like the Denver car accident victim, with rare, hard-to-match blood types.

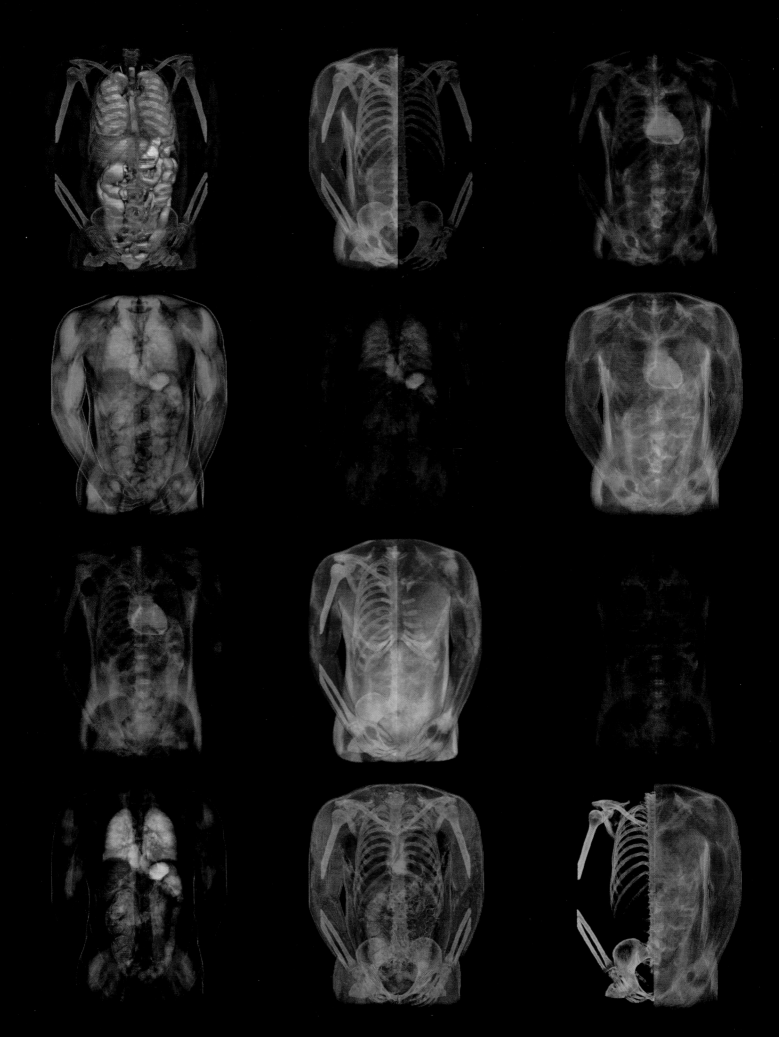

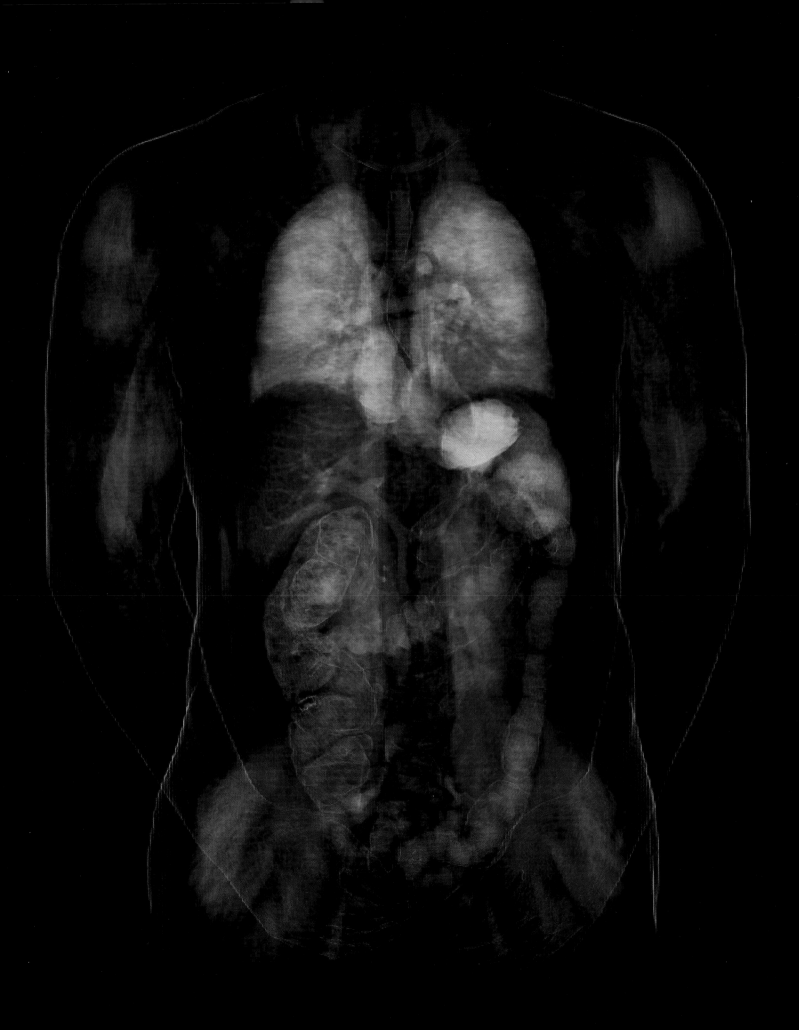

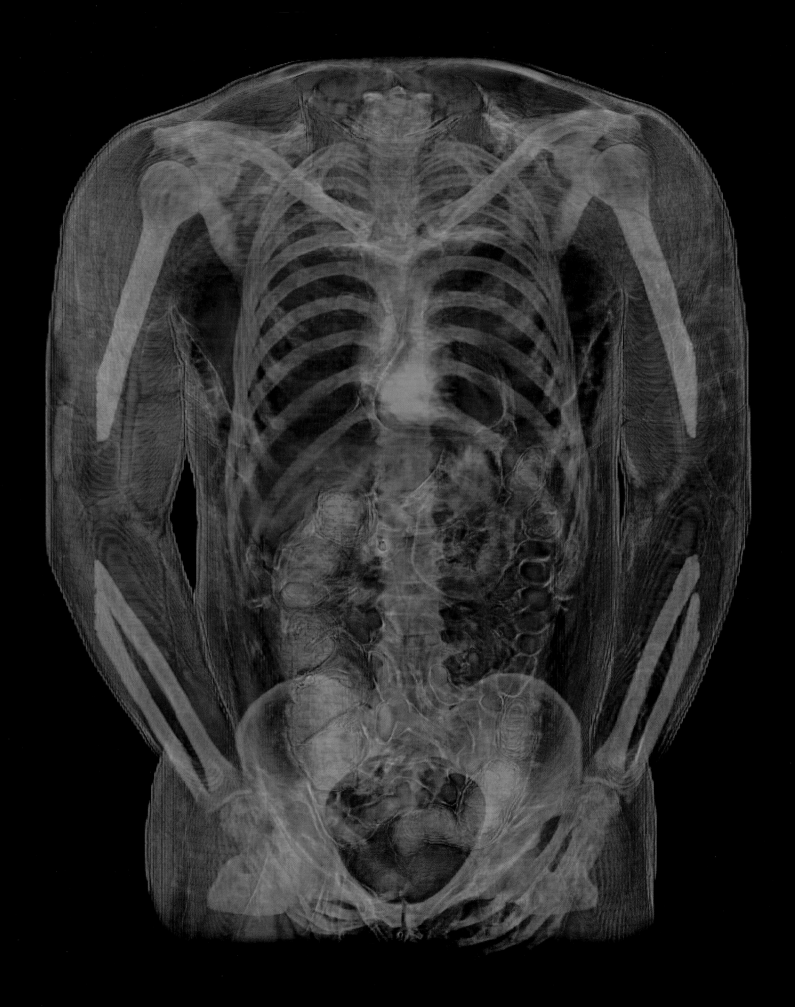

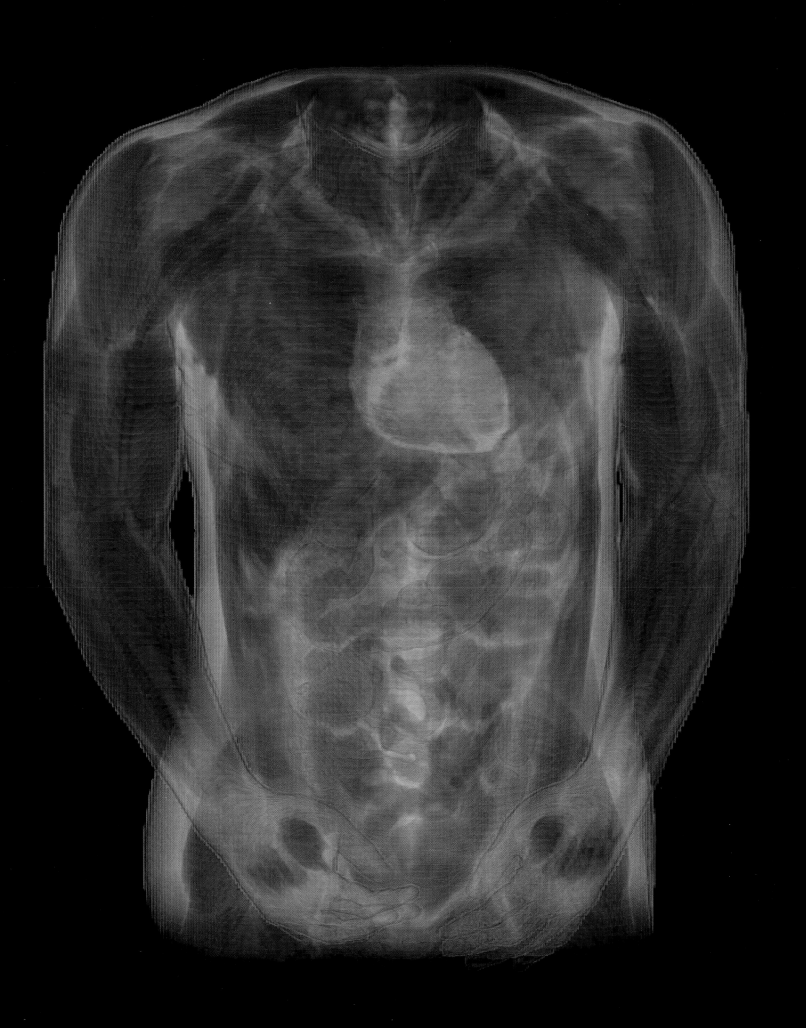

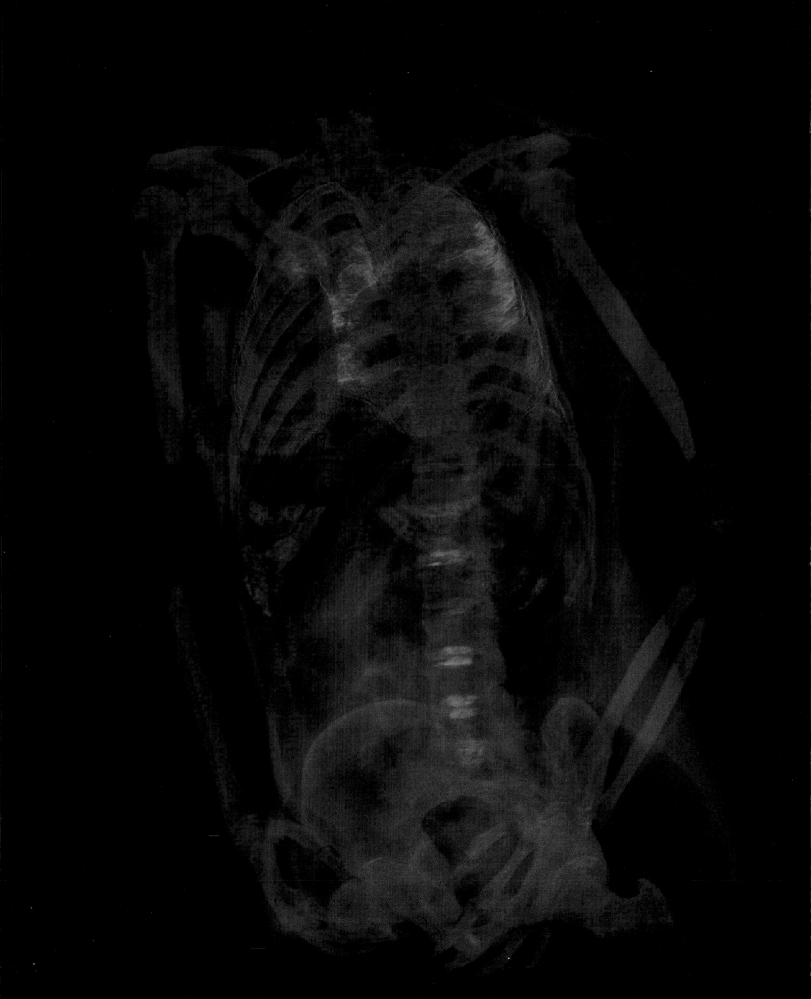

CLOSING IN ON A CURE FOR A DEADLY LUNG DISEASE:

For most 21-year-olds, life is just beginning. But for a 21-year-old with cystic fibrosis, time is running out.

There is no cure for cystic fibrosis, and patients live an average of only 28 years—long enough to fall in love, marry, raise children, and pursue careers. But their struggle to lead productive lives is waged under the shadow of frequent life-threatening hospitalizations and a constant battle for breath itself.

Cystic fibrosis strikes 1 in every 3,000 children, making it the most common fatal genetic illness. The disease is caused by a flaw in a gene that codes for a vital protein in the lining of the bronchial tubes, the airways that fan out like the branches of two inverted trees within the lungs. Lacking this natural substance, the lungs clog with thick mucus, becoming permanently damaged and leaving patients mortally susceptible to respiratory infections. Even a cold or flu can precipitate a medical crisis. But in a remarkable twist of fate made possible by contemporary biotechnology, researchers are now attempting to transform common cold viruses—normally a deadly threat to cystic fibrosis patients—into an ally in the fight to make them well. As part of a new "gene" therapy, scientists at the National Heart, Lung, and Blood Institute and elsewhere have been testing cold viruses that have been genetically engineered to carry the code for the key protein that is missing from the lungs of cystic fibrosis patients. To prevent respiratory infection, another genetic modification disables the viruses.

Once sprayed into the lungs with an aerosol inhaler or using a long, thin tube called a bronchoscope, the modified virus spreads. But instead of causing illness, it implants the missing gene into the cells lining the airways of cystic fibrosis patients, enabling them to manufacture their own supply of protein. Researchers, who have tested the therapy in over 1,000 patients, are encouraged by the results so far.

KIDNEYS

LUNG

ESOPHAGUS

ARTERIES

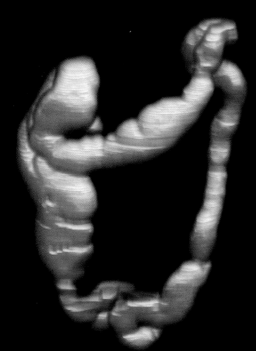

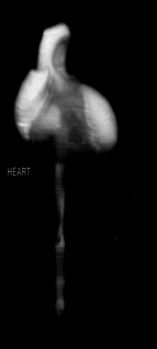

HEART

LARGE INTESTINE

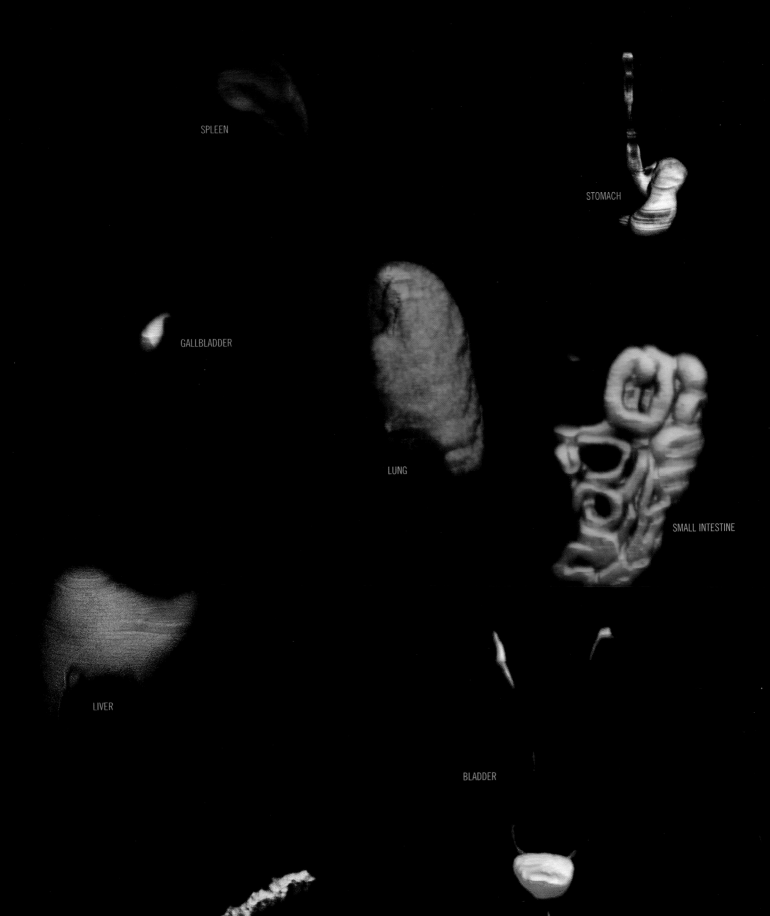

COLD-BLOODED MIRACLES:

On a winter morning in 1991, a 3-year-old girl, Brittany Eichelberger, opened the door of her parents' mobile home in Elkins, West Virginia, and walked out into a snowstorm. Three hours later, she was found unconscious and clinically dead. Her body temperature had dropped to 74°F, she had stopped breathing, and she was in cardiac arrest.

By quickly administering modern resuscitation techniques, which rapidly warm the blood by filtering it through a mechanical heat exchanger, emergency room doctors were able to pull Brittany back from the brink of death. But survival in such cases also depends on the body's own self-defense system. A fall in body temperature to below 90°F—a condition called hypothermia—constricts blood vessels, including the large left and right common carotid arteries that carry blood from the heart to the brain. As blood flow decreases, the victim becomes drowsy and loses consciousness; breathing and heart rate slow down and can ultimately stop. However, by shunting oxygen-rich blood away from the outer extremities and concentrating it in the brain and other vital organs, this natural heat- and energy-conserving blood vessel response preserves the body in a state of suspended animation—temporarily. If emergency measures can be taken fast enough—doctors have about 90 minutes before the brain and other organs begin to deteriorate—victims of life-threatening hypothermia, like Brittany, can be brought back to life, often without lasting ill effects.

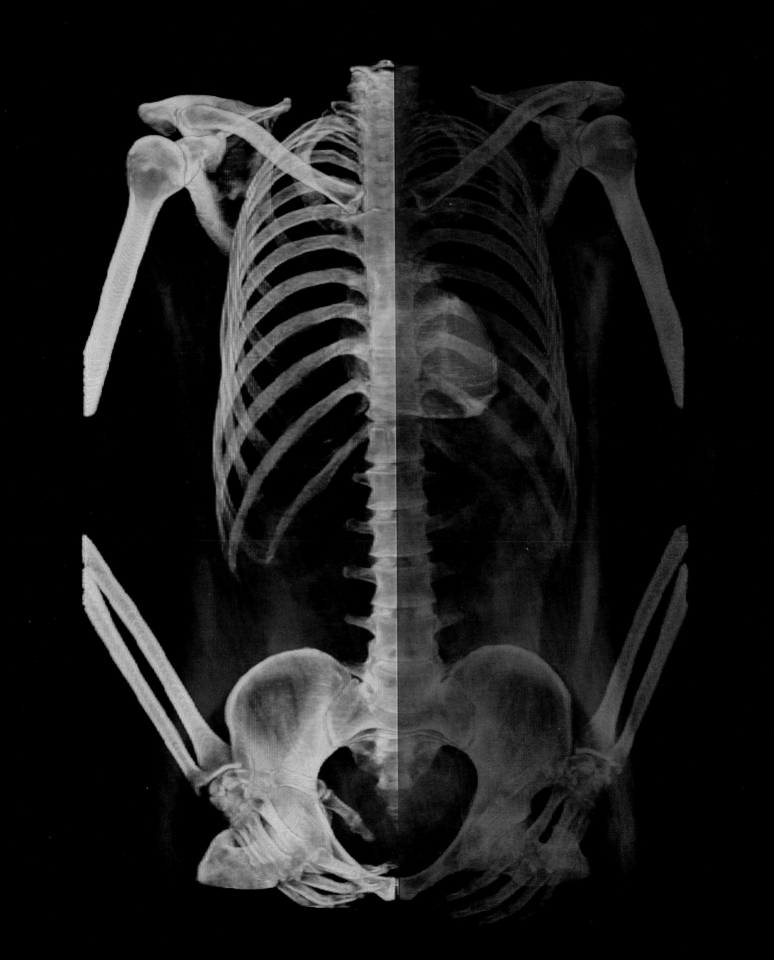

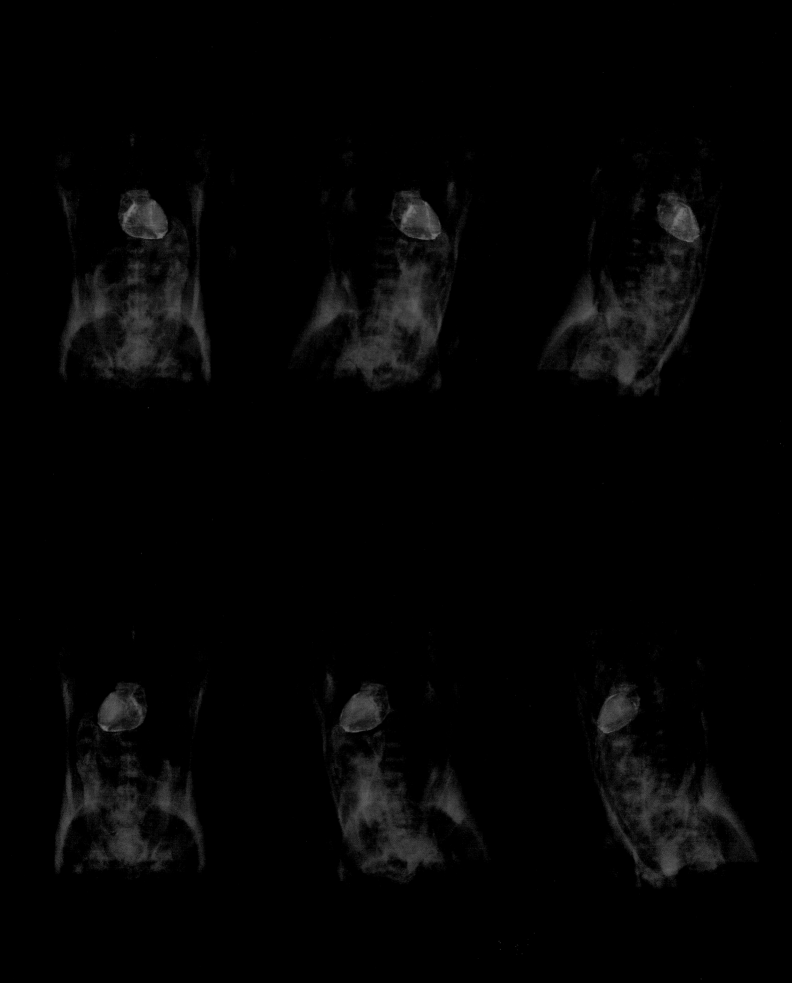

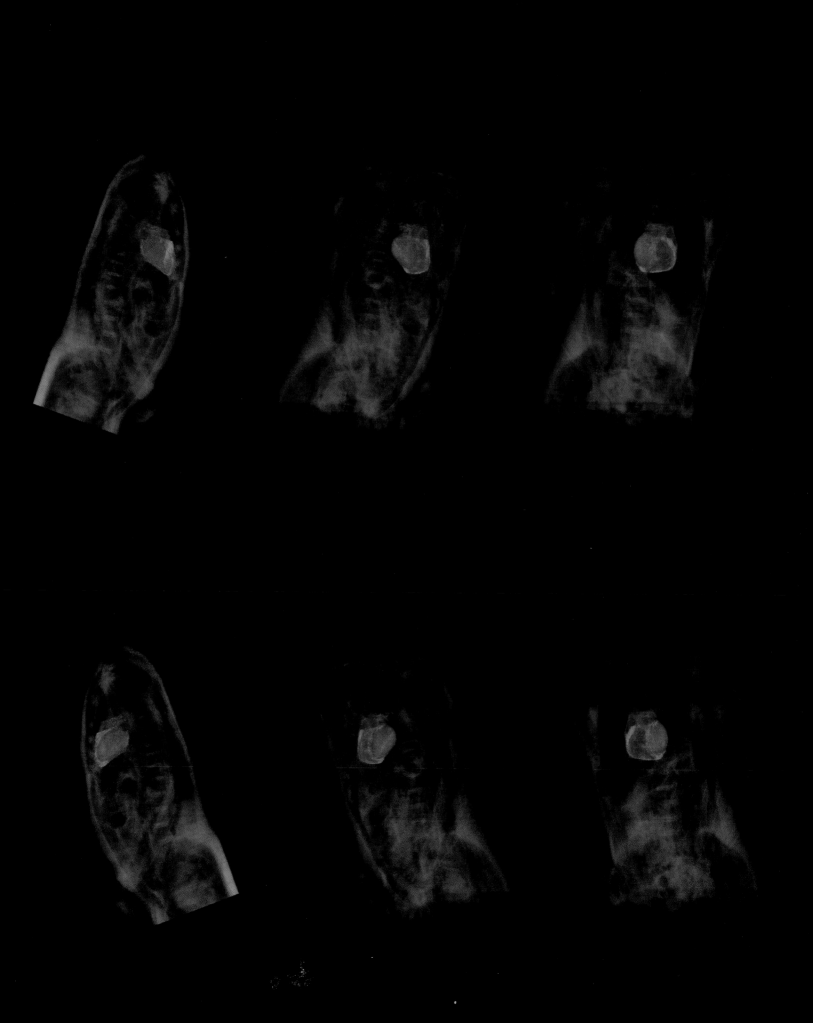

IN SEARCH OF THE MOTHER OF ALL BLOOD CELLS:

Scientists have long suspected that a small number of highly specialized parent cells are responsible for generating the trillions of cells that compose the 10 pints of blood that course through the body's 600,000 miles of blood vessels at any given moment. These hematopoietic (blood-forming) stem cells replenish the circulation not only with oxygen-carrying red cells but also with platelets, which cause blood to clot at sites of injury, and with six categories of infection-fighting white cells—a total of over 260 billion new cells every day.

Researchers know where these stem cells reside—primarily in the marrow of the upper arm, breast bone, ribs, spine, and other bones in the central part of the body. But because the blood-forming cells are so rare—only 1 in 2,000 bone marrow cells—identifying them has been a difficult task.

That hasn't stopped the dozens of scientists around the world who are racing to isolate stem cells and make them multiply, in hopes of one day being able to rebuild an entire blood supply from just a few cells. Injections of purified stem cells could be used to treat leukemia and other deadly blood disorders, rescue cancer patients whose own bone marrow has been damaged by radiation and chemotherapy, or help AIDS patients fight infection.

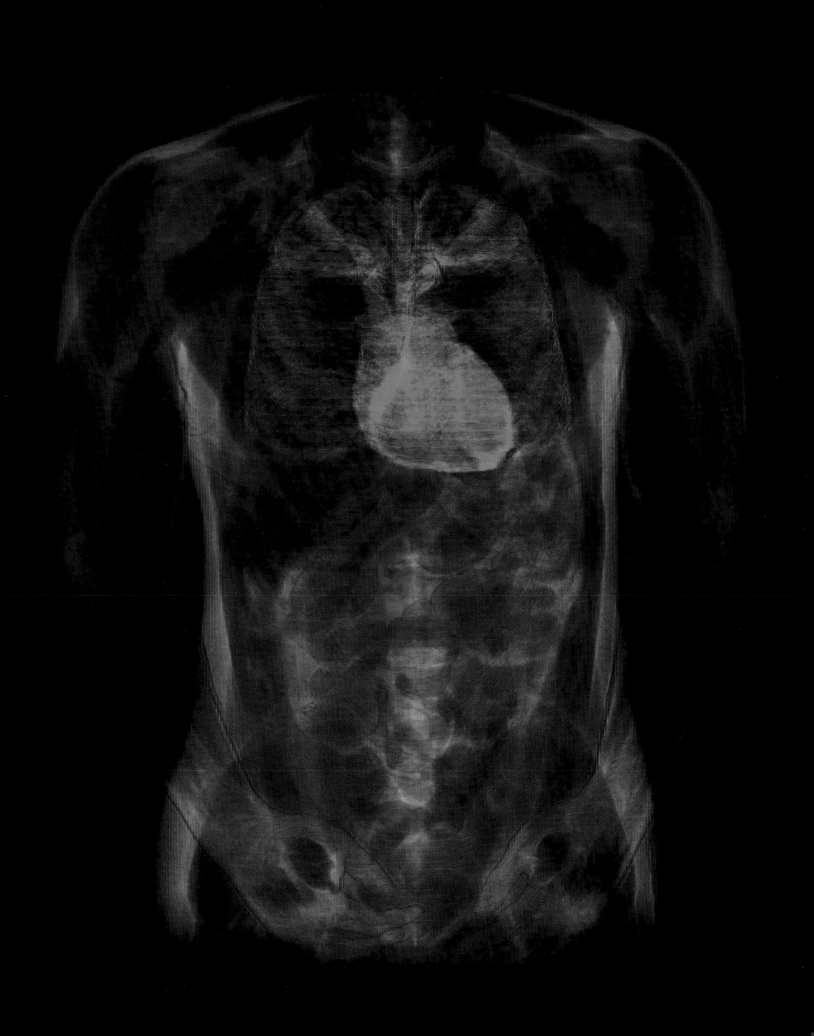

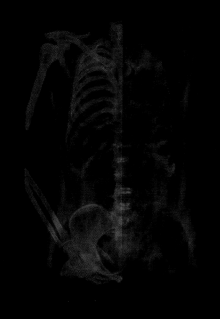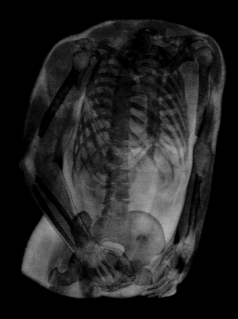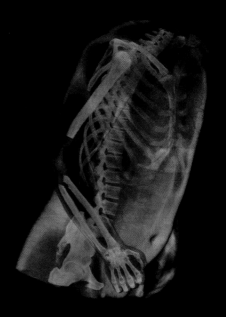
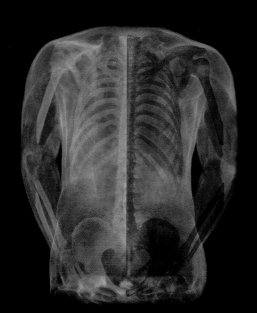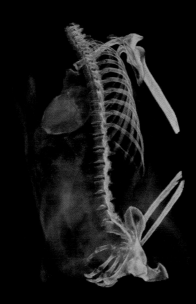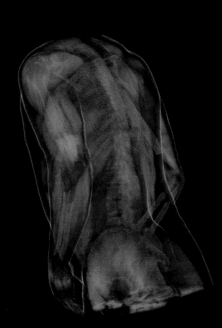

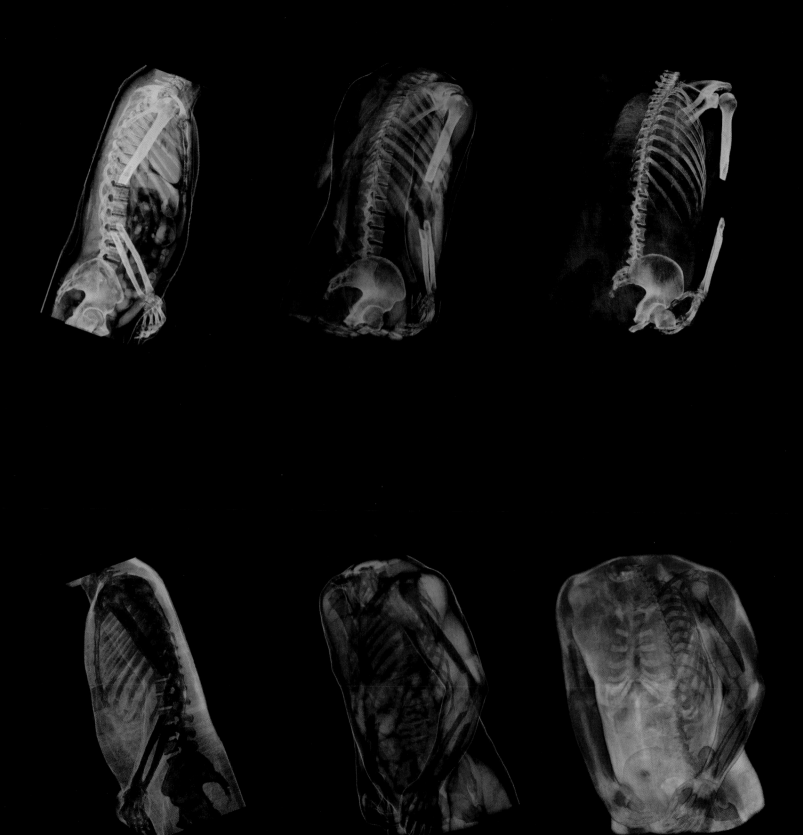

DESIGNER HEARTS:

When the Grammy Award—winning blues guitarist Johnny Copeland stepped on stage at Manny's Car Wash in Manhattan last year, he entered the annals of medical history. Copeland, who suffers from a disease called cardiomyopathy that enlarges and weakens the heart, was waiting for a transplant 6 months earlier, when his own heart failed. Without a donor heart available, doctors at Columbia-Presbyterian Medical Center in New York provided Copeland with the next best thing. They implanted a mechanical heart pump, called a left ventricular assist system (LVAS), into his abdomen just below the diaphragm and connected it with plastic tubes to his weakened heart.

Powered by a 2-pound battery pack worn on a shoulder strap, the device, which is about the size of a small tin of candy, is designed to take over the pumping function of the heart's left ventricle—for months if necessary—while enabling patients like Johnny Copeland to live close to normal lives as they wait for a new heart.

A total artificial heart (TAH) may not be far behind. Research teams in the United States and throughout the world are racing to perfect a miniature, electrically powered, implantable heart by the year 2000. Animal experiments are already under way to test several prototypes, designed to duplicate the dual-action pumping of the right ventricle, which sends blood to the lungs, as well as the left ventricle, which pumps the oxygenated blood to the rest of the body. Researchers working on the TAH admit that it will be a formidable challenge to reinvent the healthy heart's ability to pump 2,000 gallons of blood a day, year in and year out, and to dramatically vary the volume and rate of blood flow in response to the body's needs at any given moment. But if they are successful, it is estimated that in the United States alone—where the demand for heart transplants currently exceeds the supply by about 10 to 1—up to 40,000 people a year may someday have mechanical cardiac support devices like the LVAS or total artificial hearts beating inside their chests.

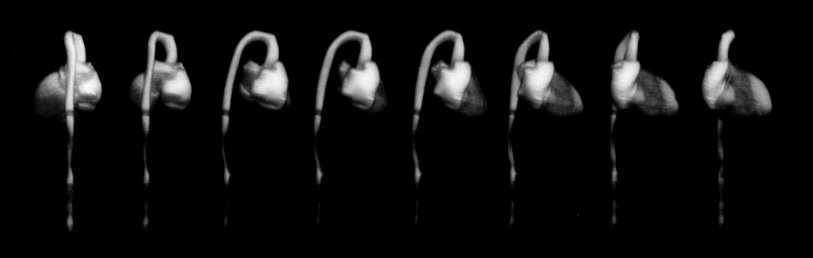

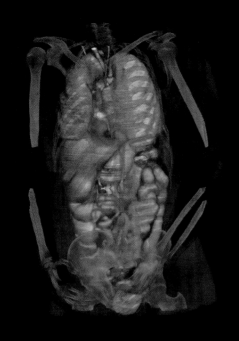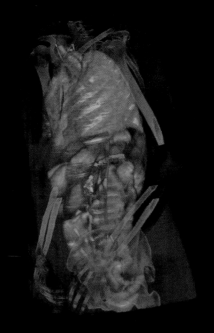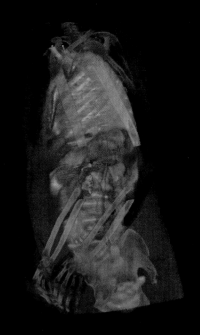
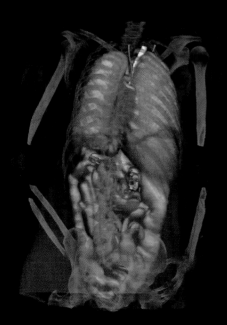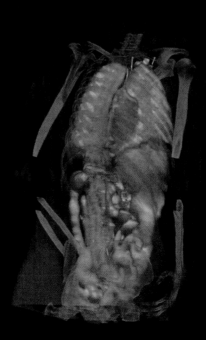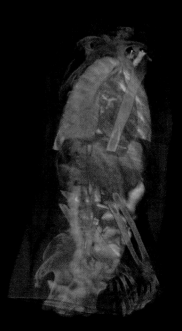

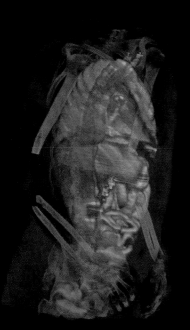
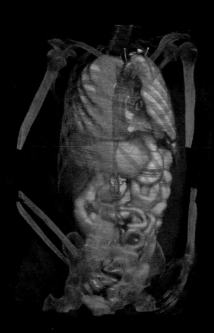
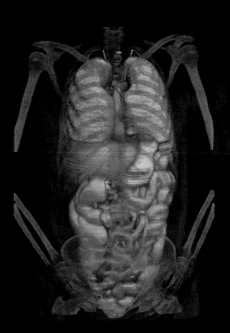
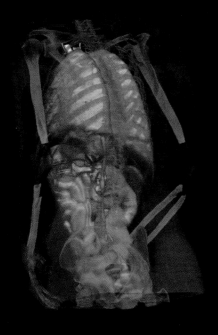
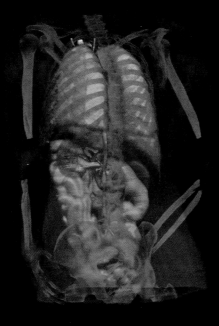
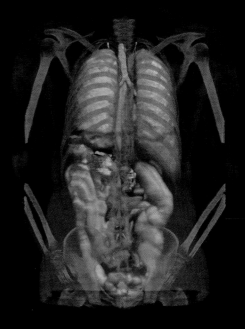

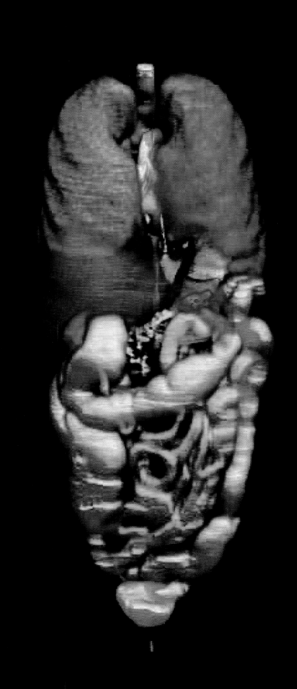

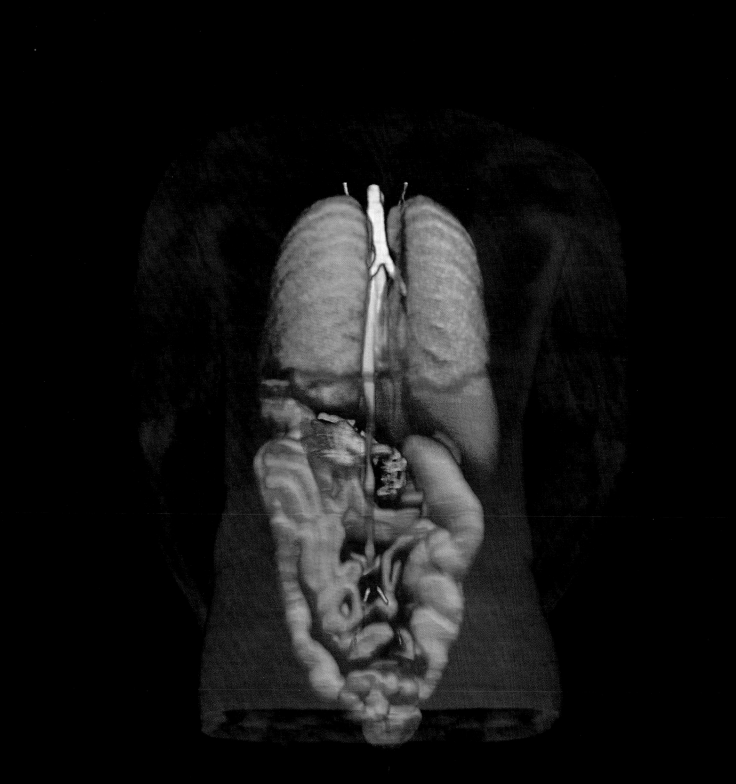

THE ULCER BUG:

Chance discoveries have played an important role in the advancement of science. But in 1983, when two Australians, Barry Marshall and Robin Warren, working in obscurity in Perth, first reported that most stomach ulcers might be caused by a bacteria, Helicobacter pylori, curable with antimicrobial drugs, most of their colleagues were skeptical. The prevailing view was that ulcers were caused by emotional stress, poor diet, and excess stomach acid.

Marshall countered his critics with a dramatic demonstration: He swallowed a beaker of liquid containing H. pylori and moments later developed an excruciating attack of gastritis. But the real proof that the microbe causes ulcers came in 1994, when the National Institutes of Health announced that the results of several large studies confirmed the Australians' original suspicions. Among ulcer patients in whom H. pylori was eliminated with antimicrobial drugs, the recurrence rate was less than 5 percent. Those who received standard therapy—drugs to block acid production—had a 75 percent rate of recurrence. So convincing is the evidence that public health experts are hopeful that antimicrobials, or possibly a vaccine, could potentially eliminate most ulcers in the United States, if not the world, in the twenty-first century.

There is still uncertainty over how H. pylori is transmitted. But the corkscrew-shaped microbe is thought to cause inflammation and open sores, both in the lining of the stomach (gastric ulcer) and in the upper part of the small intestine (duodenal ulcer), by using spidery outgrowths called flagella to penetrate the mucous layer of the stomach. The assault exposes the underlying submucosa, which is rich in nerves and blood vessels, to stomach acid and other gastric juices, worsening pain and irritation.

A STOMACH PAIN BY ANY OTHER NAME:

Physicians know that when a patient complains of a "stomachache," the actual cause may be any number of conditions affecting the organs packed into the abdomen. While some bellyaches may be due to stomach ailments, such as gastritis or ulcers, the possible explanations also include hernias, gallstones, kidney stones, or urinary tract infections, bowel disorders, and, in women, a host of problems that affect the reproductive organs. To begin to make a diagnosis, doctors usually mentally divide the patient's abdomen into four quadrants, by drawing imaginary lines—one vertical, one horizontal—through the navel. Each section contains specific organs. The right upper quadrant contains the liver, the gallbladder, the duodenum of the bowel, and part of the pancreas; the left upper quadrant: the spleen, colon, stomach, and parts of the pancreas and lungs. Pain in the left lower quadrant narrows down the problem to an intestinal virus, kidney stones, or a bowel disorder. Pain starting around the navel and radiating down to the right lower quadrant is a classic symptom of appendicitis.

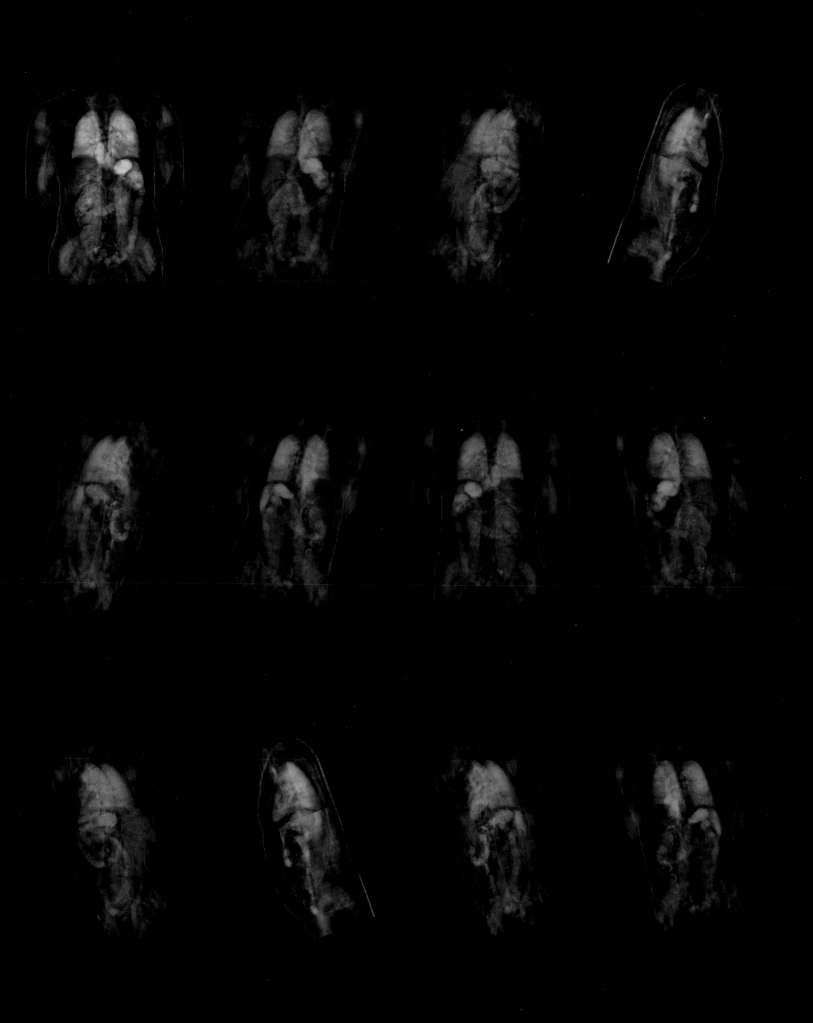

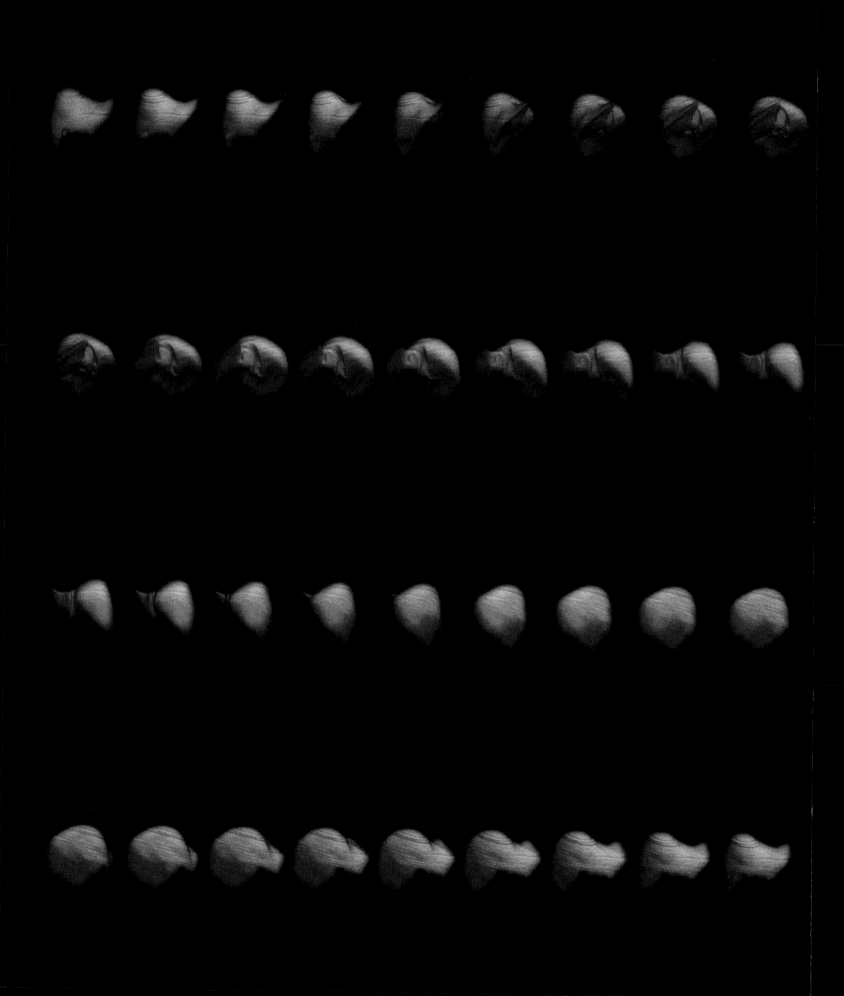

WHO SHALL LIVE AND WHO SHALL DIE?

In June, 1995, when Mickey Mantle underwent a transplant operation to replace a liver ravaged by cancer, hepatitis, and cirrhosis, his doctors could only give him a slightly better than 50 percent chance of living for 3 years. Without a transplant, the baseball legend's chances of survival were zero. But were the odds good enough to justify $300,000 in medical expenses and a scarce donor liver in the effort to save Mantle—who died 2 months later from a recurrence of cancer?

For many patients with advanced liver disease, the benefits do outweigh the risks. With improved surgical techniques and antirejection drugs, four out of five liver transplant recipients now live 5 years or longer with healthy, new organs replacing their own failing liver's ability to produce life-sustaining enzymes and chemicals and to filter the blood of waste products. But the operation and follow-up care are costly—averaging $250,000—and the demand for donor organs exceeds the supply by more than two to one. Who should get the six or seven precious organs that become available each day?

In Great Britain and other countries, Mantle's age, 63, might have eliminated him as a transplant candidate because donor organs are restricted to younger patients, who are thought to have a better chance of survival. Transplantation for liver cancer is also controversial because of the possibility that the disease has spread. But the operation also raised another troubling question over whether alcoholics, even recovering alcoholics like Mantle, should receive the same priority as patients whose behavior has not contributed to their disease. With alcoholic cirrhosis accounting for over 40 percent of the 26,000 deaths in the United States each year from chronic liver disease, patients with a history of alcoholism are more commonly considered as transplant candidates than they were 10 years ago, in part because of the success of substance abuse programs. But many hospitals continue to stipulate that heavy drinkers must first demonstrate that they have been alcohol-free for 6 months before placing them on the waiting list for a new liver.

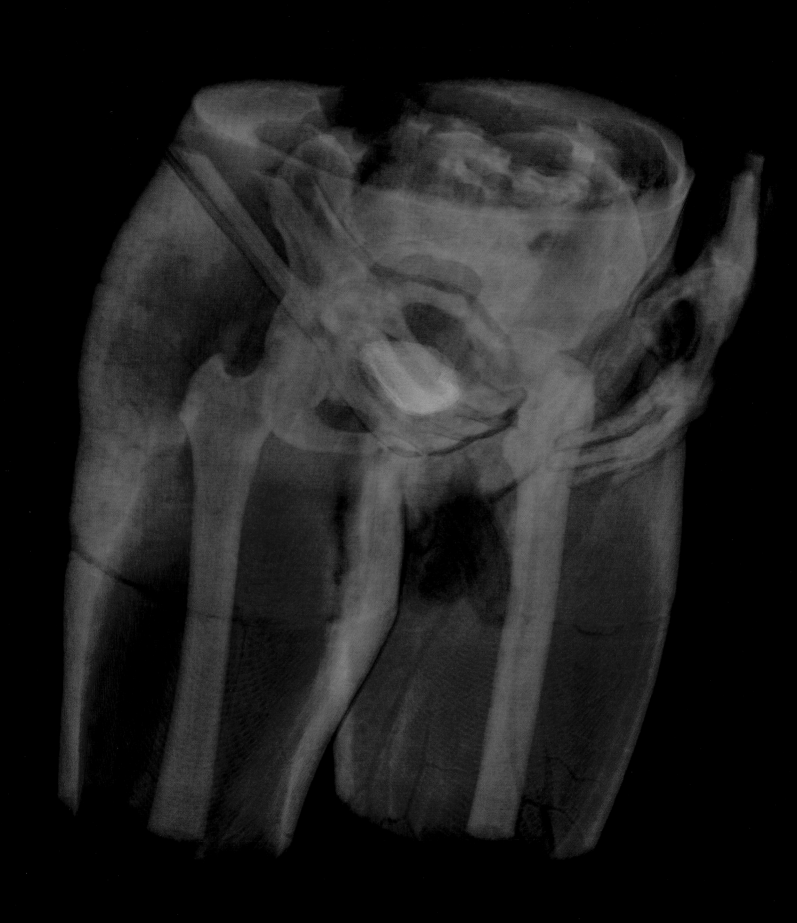

The Pelvis

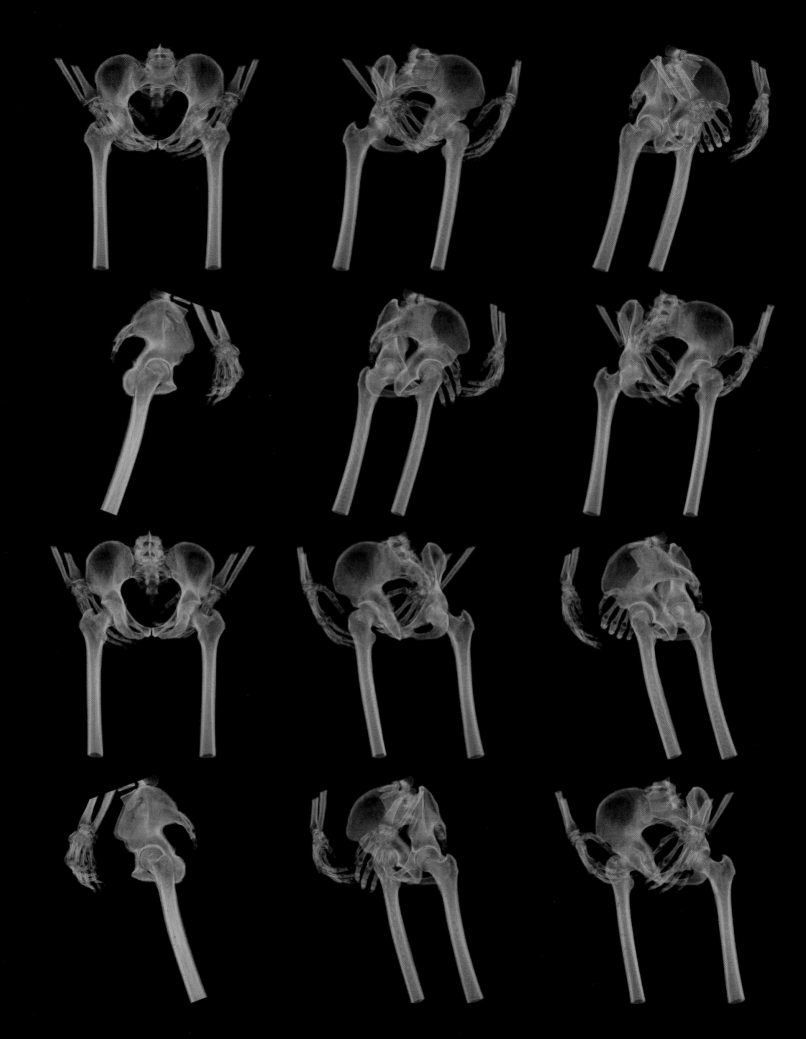

While the heart and organs of the upper body have historically been exalted as the seat of love, creativity, and robust health, the pelvis has been reviled as a kind of anatomical netherworld, the source only of excrement and lust, far removed from the higher realms of spirit and mind.

Take the sacrococcygeal region. The large Gluteus maximus muscles are, in fact, a unique hallmark of human identity, a remarkable evolutionary stroke of genius, which helped to enable us to first stand up and walk. The magnificently designed ball-and-socket joints of the hip, crafted by nature to permit our legs the wide range of movement that allows us to run, jump, and dance, have been successfully imitated by scientists working on artificial hips—but never duplicated.

The looping coils of the large and small intestine, far from being a mere 30 feet of plumbing, are a sophisticated food-processing plant for the production of the nutrients that fuel our cells.

The *underworld* of the pelvis does harbor some unusual inhabitants—some 5,000 species of bacteria live in the small intestine, performing a vital digestive role. But the pelvic region's lowly past reputation is clearly unjustified. The fact that the intestines and other digestive organs even have a "brain" of their own—a network of millions of nerve cells that not only share the basic wiring of the brain but echo its emotional states on a *gut* level—is evidence that, if anything, the body is a unified whole, not a collection of separate, unequal parts.

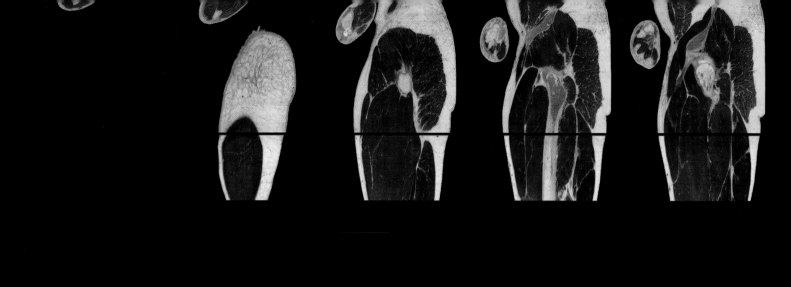
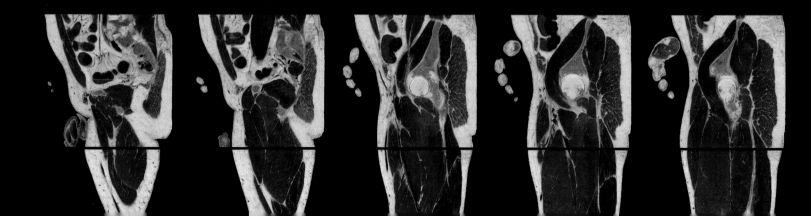

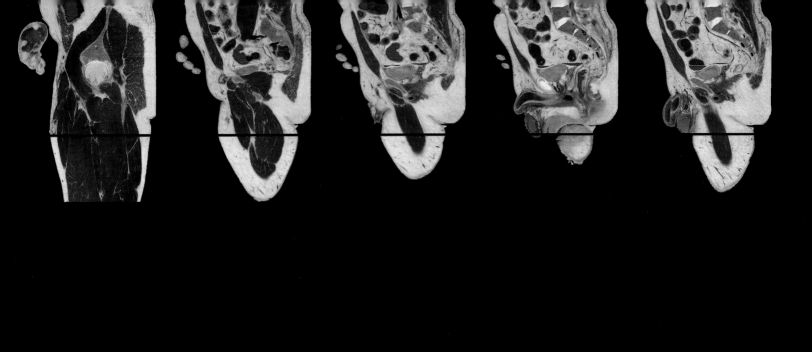

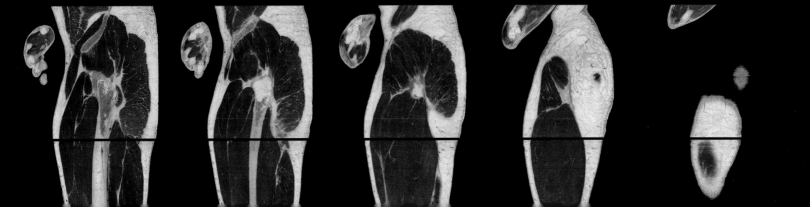

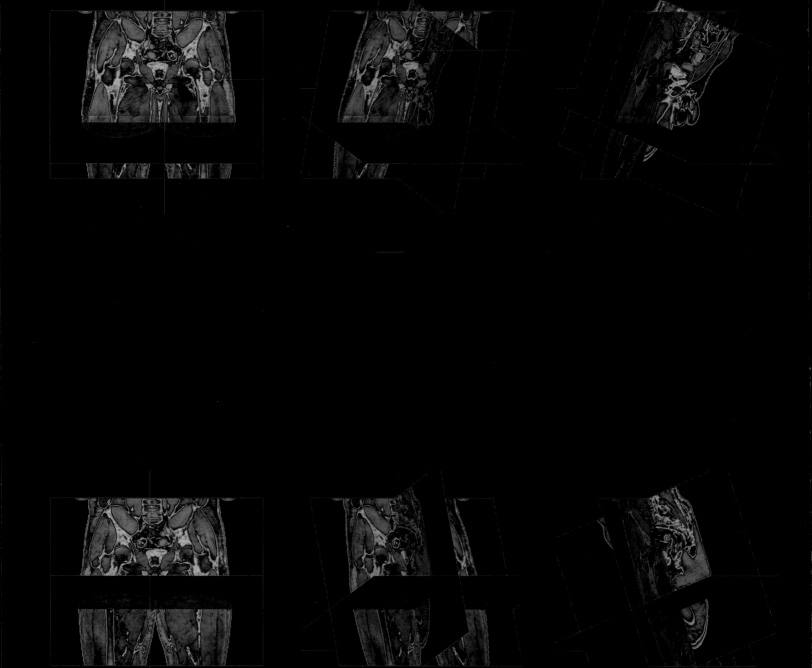

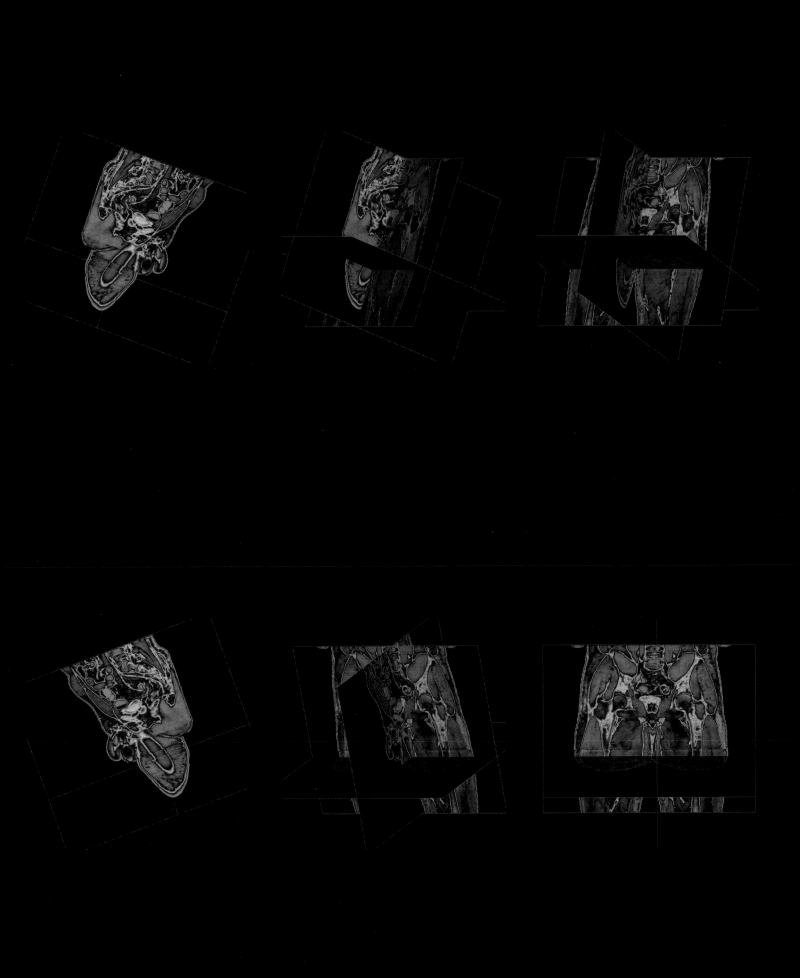

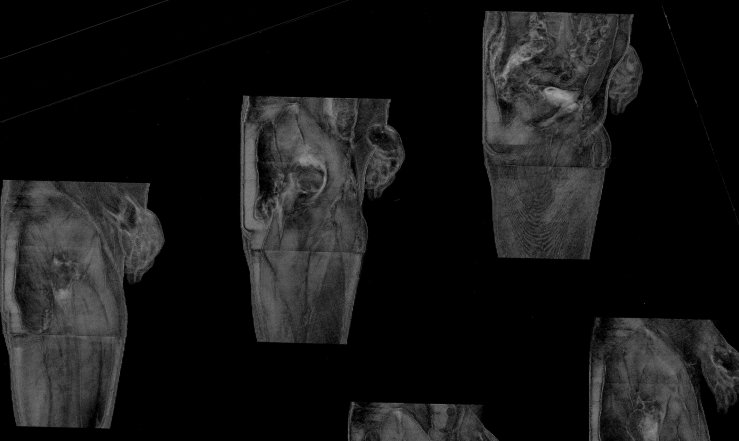

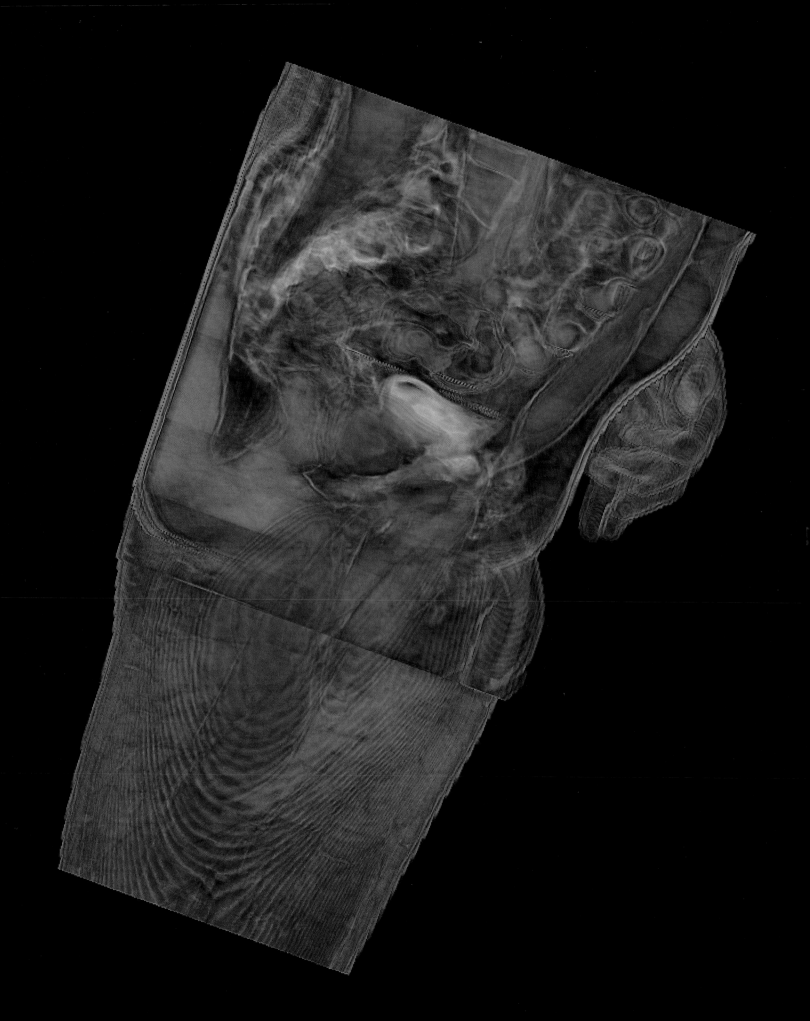

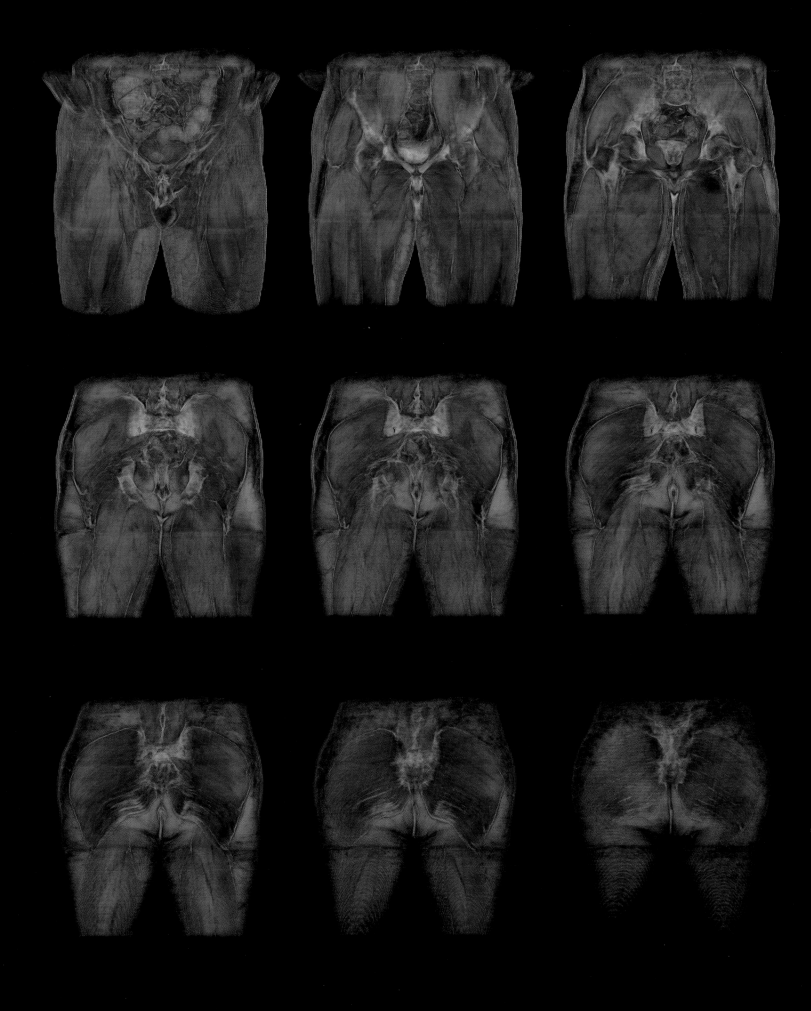

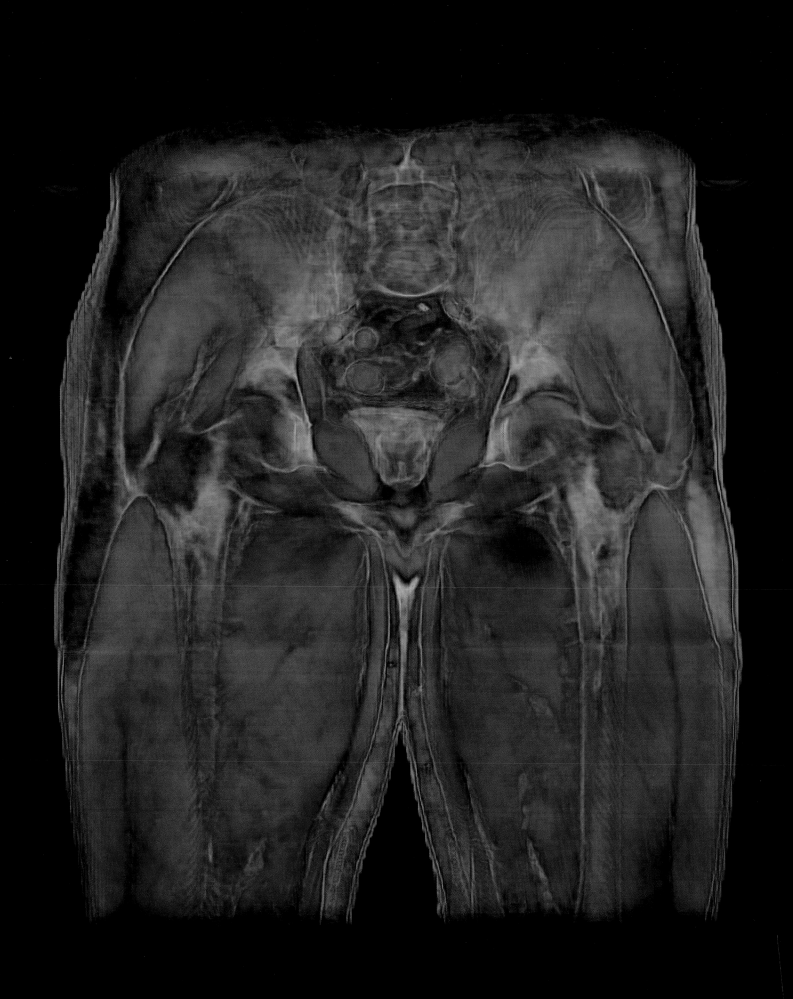

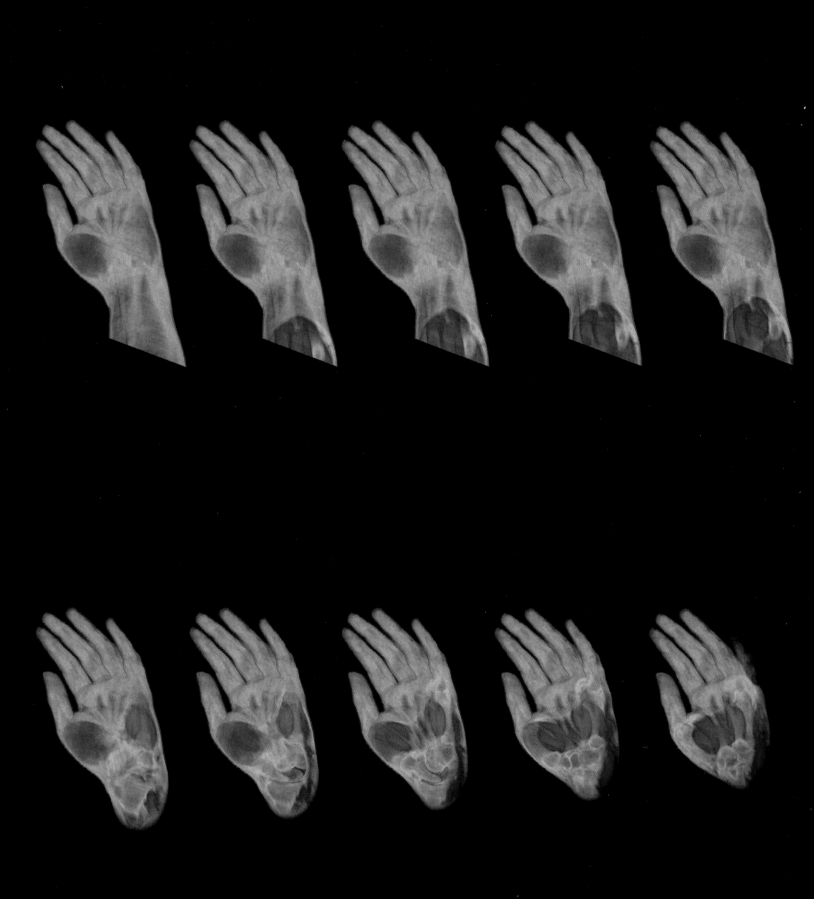

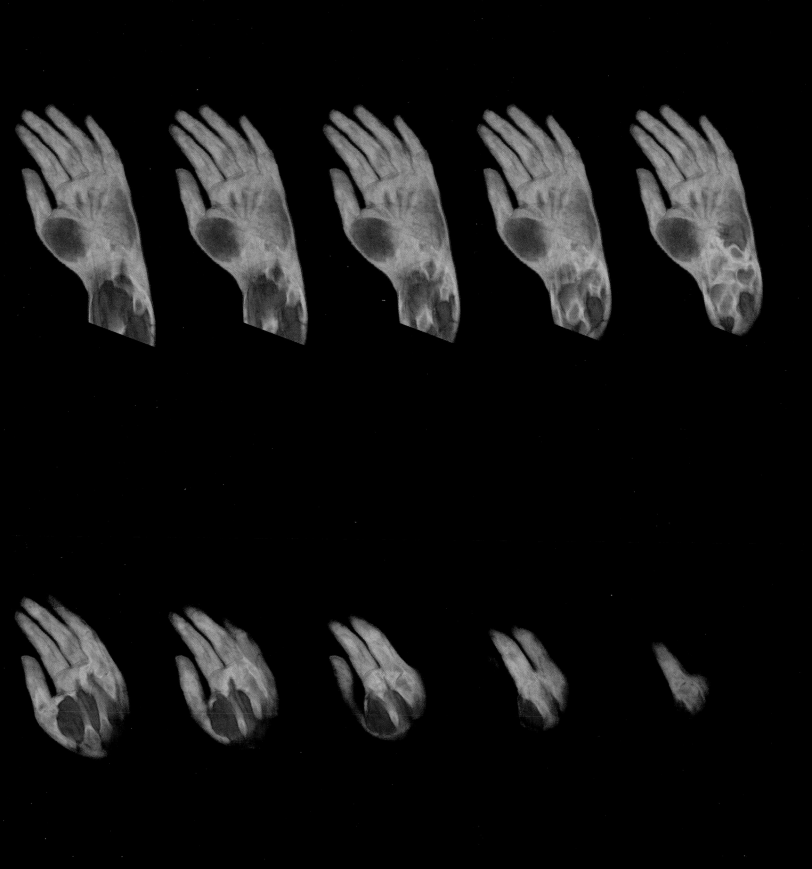

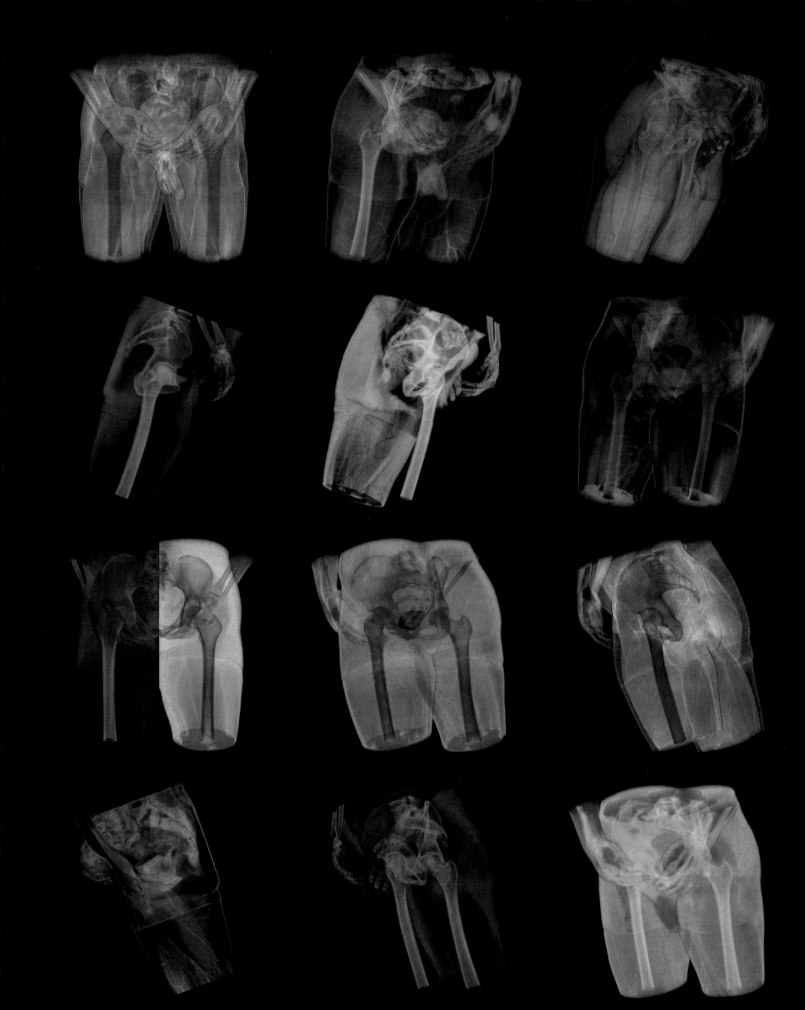

THE "AGGRESSION" HORMONE?

Not long ago, many psychologists believed that excesses of the male hormone testosterone could trigger the sort of murderous mayhem for which Joseph Paul Jernigan was executed. But is raging testosterone really to blame for male aggression? The answer, according to recent research, is no. Testosterone, which is produced along with sperm in the testicles, does exert powerful effects on men's bodies from the very beginning of life. A masculinizing hormone, or androgen, testosterone promotes the development of the penis and scrotum during gestation and the maturation of the reproductive organs during puberty, stimulates the sex drive, and contributes to numerous other common masculine traits, from baldness to the accumulation of abdominal fat—the male "spare tire." Scientists reasoned that because injections of high doses of testosterone produced attack as well as mating behavior in both male and female laboratory animals, it might also be the dreaded aggression hormone. However, human evidence is lacking. In fact, just the opposite may be true. Some studies have associated violent responses in animals and humans to an overabundance of the female hormone estrogen, which, like testosterone, is shared by both sexes in differing ratios. Other research has found that men with testosterone deficiencies who were undergoing hormone replacement therapy scored higher for anger, irritability, and aggressive behavior before their shots, when their testosterone levels were low. However, low testosterone is no excuse for murder in Jernigan's case. His missing testicle was removed due to a benign tumor while he was in prison

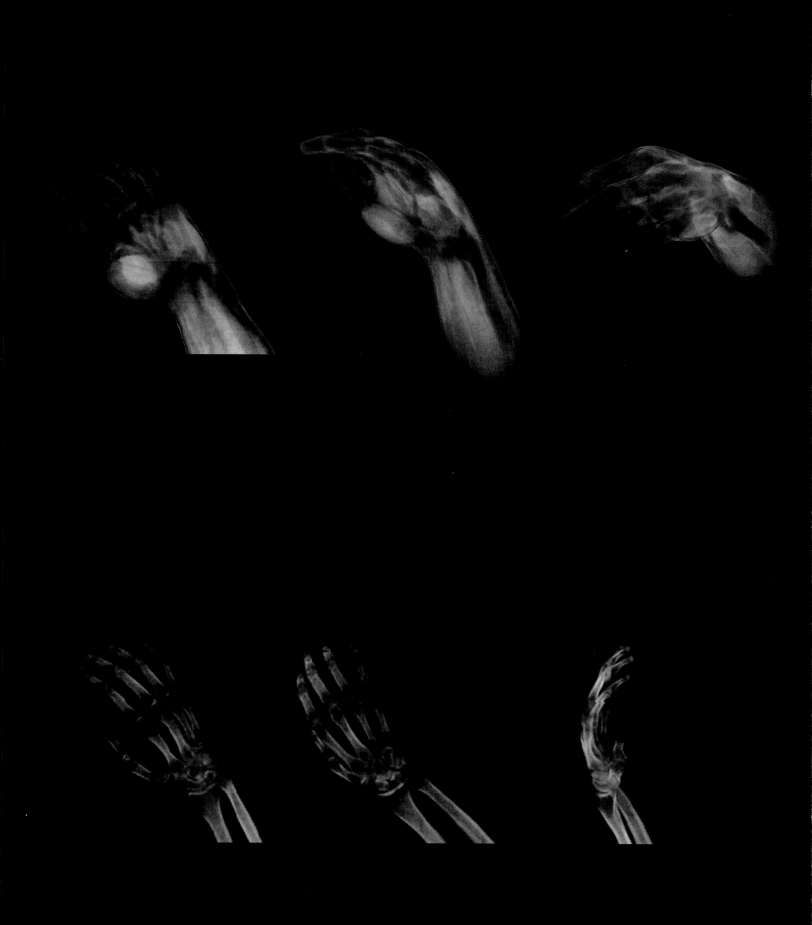

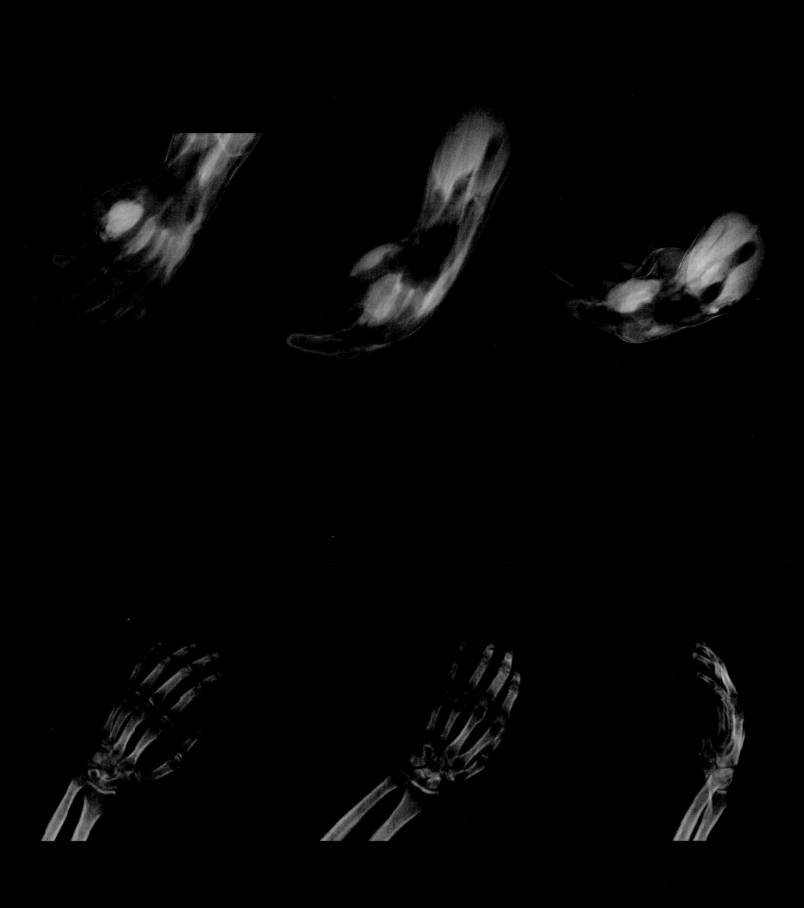

GUT FEELINGS:

"Butterflies" in the stomach and nervous diarrhea are the responses to emotionally stressful situations of a second brain, located not in the skull but in the gut. Called the enteric nervous system, the gut's brain is a complex, interconnected network of 100 million nerve cells—10 times more than are present in the spinal cord—located in sheaths of tissue lining the esophagus, stomach, small intestine, and colon. The two brains are linked by the vagus nerve, which descends to the gut from the brain stem, and use the same neurotransmitters to send chemical messages back and forth among nerve cells. The enteric nervous system's function is mainly to sense and communicate information about levels of sugar, protein, acidity, and other chemical factors that coordinate the muscle contractions of the gut during digestion. But because of the built-in connections and physiological similarities, the brain's reactions to emotional states are mirrored in the gut as well. A surge of neurotransmitters and hormones originating in the brain in response to stress or fear and cascading through the enteric nervous system can trigger muscle spasms in the colon, causing diarrhea, or constriction of muscles in the stomach or esophagus, resulting in the sensation of "butterflies" or being "choked" with emotion. Another striking relationship between the two brains occurs during sleep. Researchers have found that at night the gut's brain produces 90-minute cycles of slow-wave muscle contractions punctuated by short bursts of rapid firing, similar to the brain-wave patterns associated with dreaming.

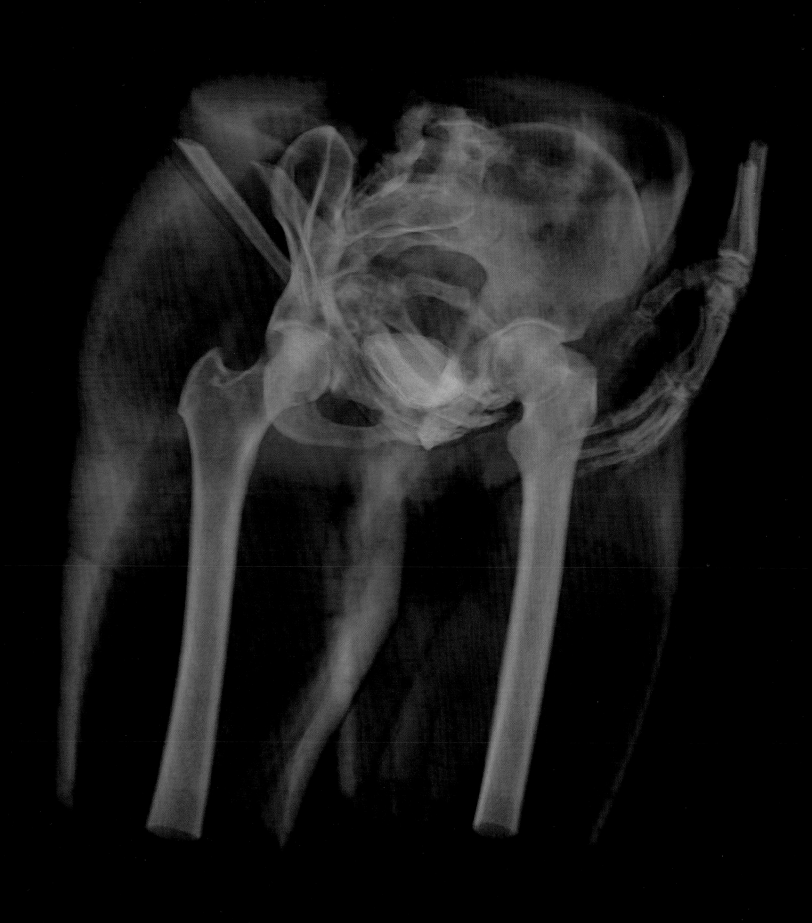

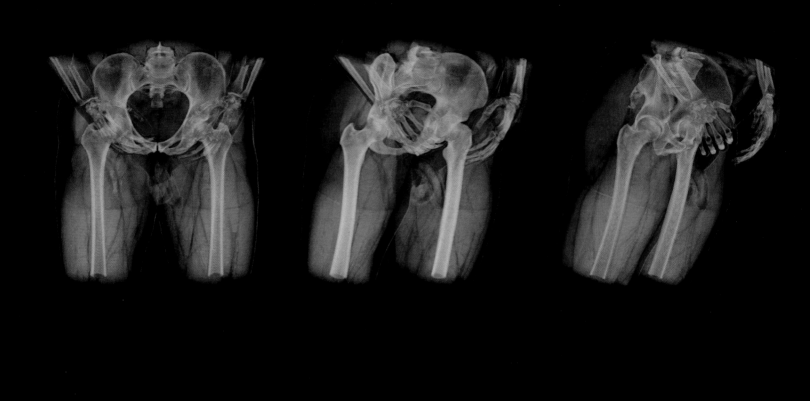
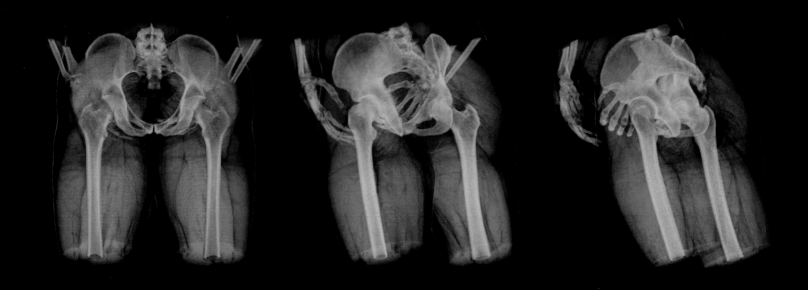

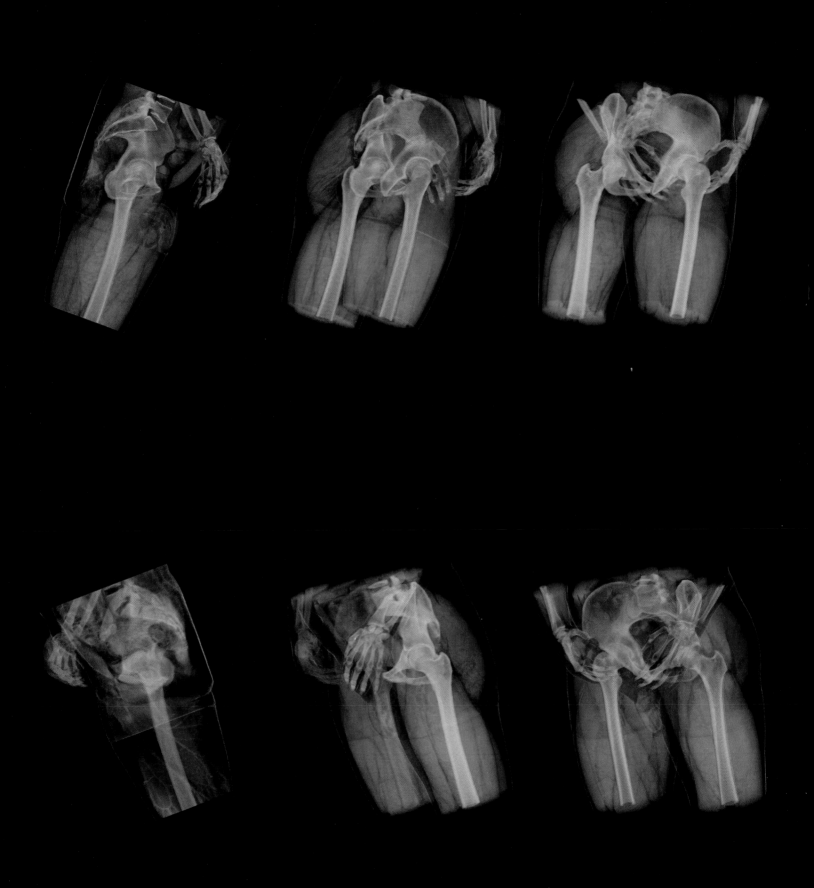

THE CIRCUMCISION CONTROVERSY:

"My own preference, if I had the good fortune to have another son, would be to leave his little penis alone," America's best-known baby doctor, Benjamin Spock, wrote in 1989.

Medical opinion regarding circumcision—the removal of the foreskin of a newborn's penis—has fluctuated dramatically in the past 30 years. Practiced ritually by Jews and Muslims for 4,000 years, circumcision began to be more widely accepted late in the last century in the belief that it promoted hygiene and prevented venereal diseases. By the 1960s, fully 90 percent of all male newborns in the United States were being circumcised as a routine procedure. (Joseph Paul Jernigan, born in 1955, was an obvious exception.) In early editions of his famous book *Baby and Child Care,* Spock originally agreed with the pro-circumcision position. But he changed his mind—along with many other physicians—in the early 1970s, when the American Academy of Pediatrics stated its opposition to routine circumcision, citing a lack of evidence of actual health benefits. An influential article in the *New England Journal of Medicine* appearing at that time termed the practice "ritualistic surgery." Opposed as well by insurance companies who did not want to pay for the procedure, circumcision of newborn boys dropped by one-third in the years that followed.

Recently, however, the pendulum has begun to swing in the other direction, based on new scientific evidence that circumcision reduces the risk of urinary tract infections in infant boys and virtually eliminates the threat of penile cancer later in life. Studies also indicate that uncircumcised men run a higher risk of becoming infected with the virus that causes AIDS. The reason may be that an intact foreskin provides a warm, moist environment that allows the HIV virus—which is usually short-lived outside its human host—a greater opportunity to survive and penetrate the body.

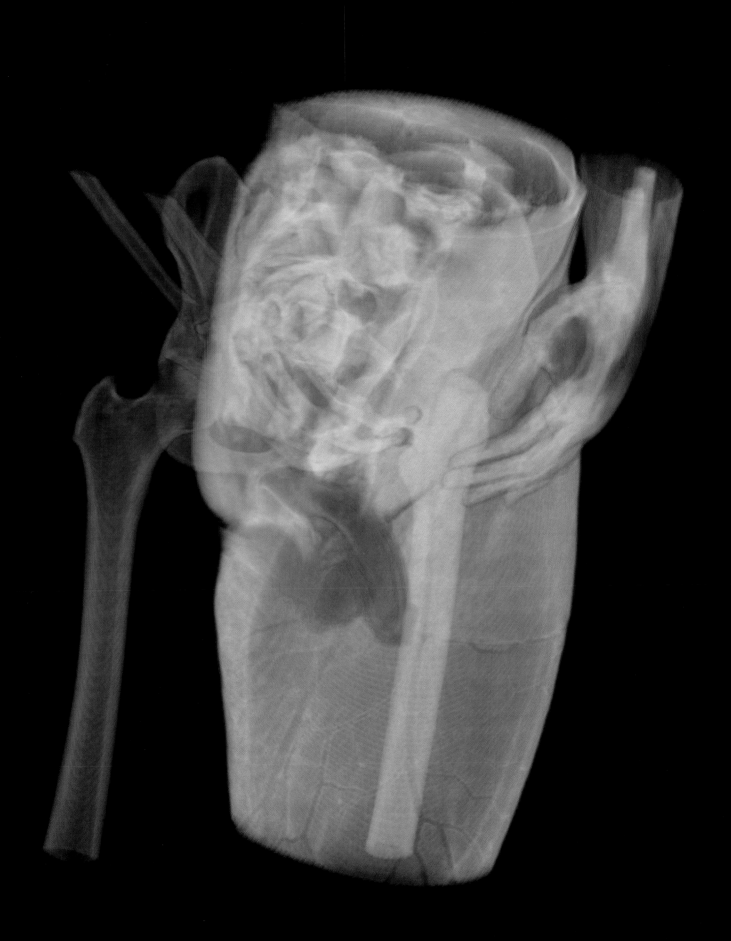

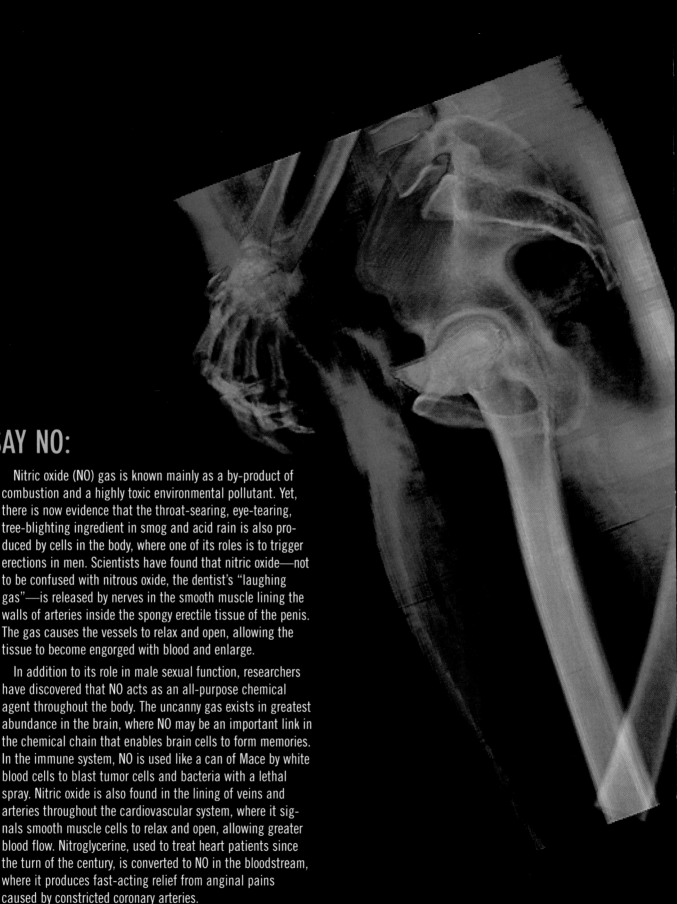

JUST SAY NO:

Nitric oxide (NO) gas is known mainly as a by-product of combustion and a highly toxic environmental pollutant. Yet, there is now evidence that the throat-searing, eye-tearing, tree-blighting ingredient in smog and acid rain is also produced by cells in the body, where one of its roles is to trigger erections in men. Scientists have found that nitric oxide—not to be confused with nitrous oxide, the dentist's "laughing gas"—is released by nerves in the smooth muscle lining the walls of arteries inside the spongy erectile tissue of the penis. The gas causes the vessels to relax and open, allowing the tissue to become engorged with blood and enlarge.

In addition to its role in male sexual function, researchers have discovered that NO acts as an all-purpose chemical agent throughout the body. The uncanny gas exists in greatest abundance in the brain, where NO may be an important link in the chemical chain that enables brain cells to form memories. In the immune system, NO is used like a can of Mace by white blood cells to blast tumor cells and bacteria with a lethal spray. Nitric oxide is also found in the lining of veins and arteries throughout the cardiovascular system, where it signals smooth muscle cells to relax and open, allowing greater blood flow. Nitroglycerine, used to treat heart patients since the turn of the century, is converted to NO in the bloodstream, where it produces fast-acting relief from anginal pains caused by constricted coronary arteries.

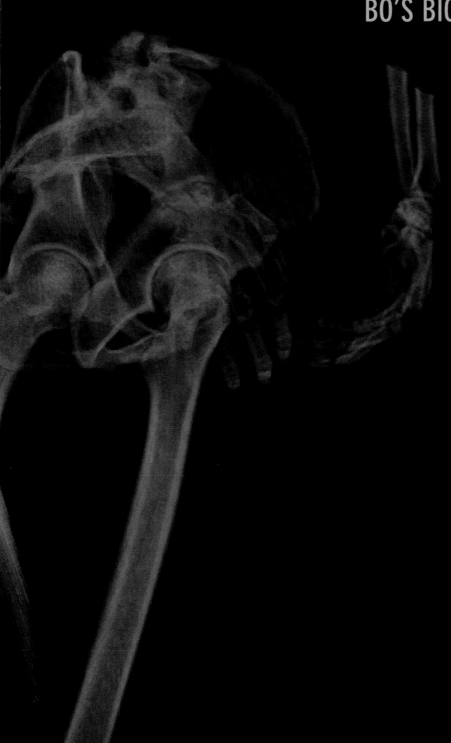

BO'S BIONIC HIPS:

Bo Jackson's 1994 baseball season was one for the medical record books. Two years earlier, the two-sport star, who gained fame as both an outfielder for the Kansas City Royals and a Los Angeles Raiders running back, suffered a severe hip injury in a National Football League playoff game against the Cincinnati Bengals. The bone-crushing tackle that ripped his thigh bone from his hip socket, stripping away the vital lining of cartilage needed to cushion the joint, should have ended Jackson's athletic career. But after hip replacement surgery, Jackson made a comeback the following summer, hitting 16 home runs for the pennant-winning Chicago White Sox, with the aid of a plastic and metal artificial left hip.

Jackson's amazing feat added fuel to a raging debate among orthopedic surgeons over who should receive hip replacements. Some 150,000 hip replacements are performed each year, mostly in older people hobbled with degenerative joint disease, with only a third of new artificial hips going to patients under age 65. In the procedure, surgeons remove the round head of the thigh bone, driving a metal spike into the marrow canal. A metal ball wedged onto the spike fits into a cup made of friction-reducing polyethylene plastic, which replaces the hip socket. Although some doctors point to patients like Bo Jackson as examples of how effectively the bionic hips can restore mobility and provide relief from pain, they are no match for the natural hip's extremely well-wrought ball-and-socket joint, which affords the leg a wide range of pain-free movement. Polymer chemists have also been unable to duplicate the tough, elastic cartilage and lubricating synovial fluid within the hip socket. As a result, many hip replacement patients continue to complain of constrained movements and residual discomfort after their operations. Nor are the implants truly permanent. The cement used to fasten the artificial joint to the skeleton often breaks down within 15 years, and wear and tear on polyethylene components can release plastic particles into the surrounding tissue, causing inflammation and bone loss. Because second replacements are more difficult and less successful, doctors have been reluctant to recommend the operation for younger patients.

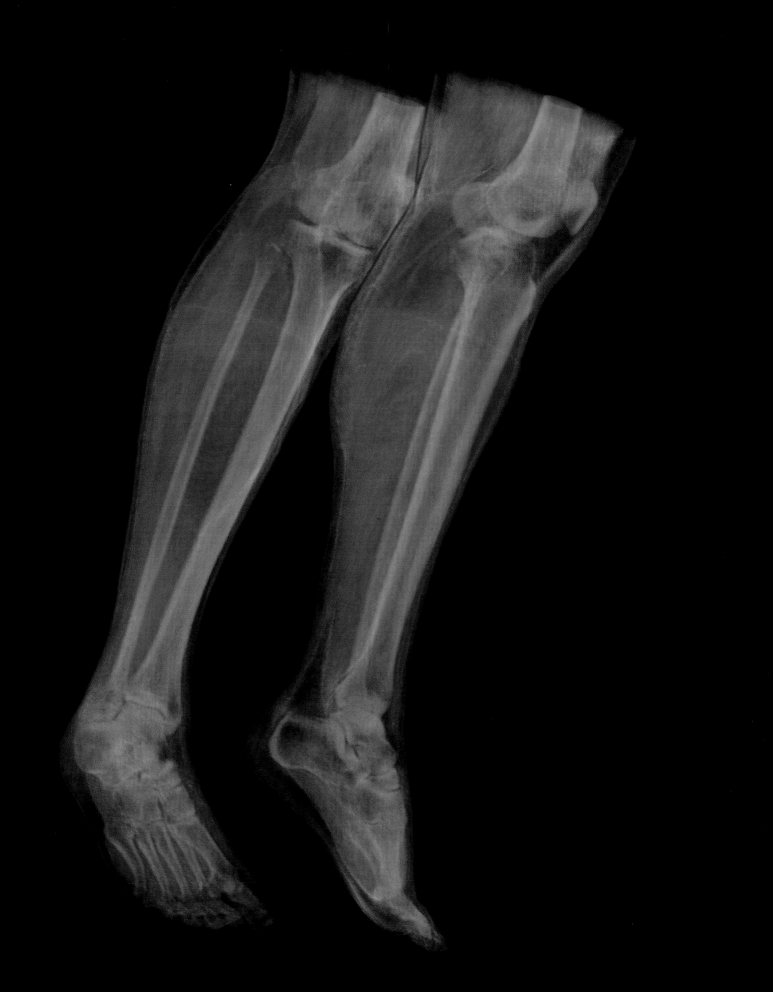

The Legs

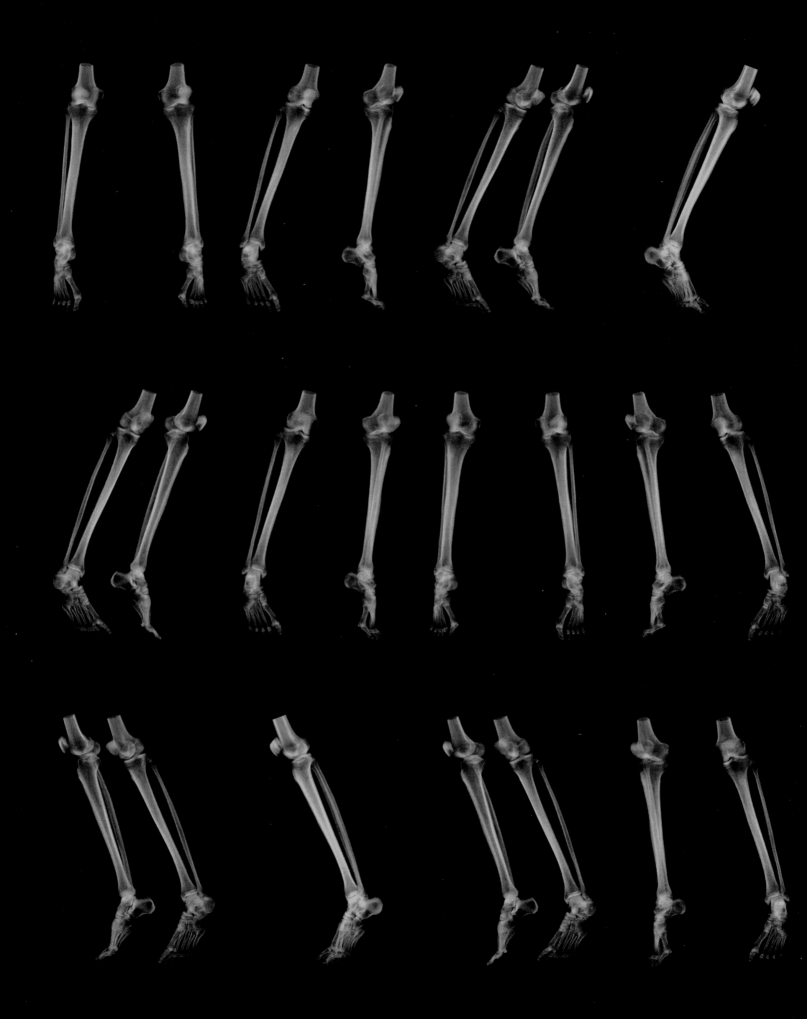

The naturalist Stephen Jay Gould has observed that the development of a large brain, long considered to be the hallmark of human evolution, was actually an easy transformation. Upright posture is the real surprise, the fundamental reconstruction of our anatomy, that really gave birth to humankind.

The distinctly human bones and muscles of the shin, thigh, knee, and foot—not a big brain—mark the separation between man and ape some 6 million years ago; we have been walking on our own two legs and feet ever since.

To do so, we have evolved intricate muscle and skeletal structures—the ingenious structure of the foot with its stabilizing big toe can absorb up to 20,000 pounds—of pressure and a nervous system capable of making the split-second connections to the brain that are needed for balance and complex movements. These same nerve networks also contribute to the strange phenomenon of "phantom" limbs after patients lose a leg or arm.

Standing upright has proven to be extremely advantageous, even though anthropologists still do not know why human evolution favored a five-toed foot, or chose an upright posture—the most unstable mode of locomotion possible. Upright movement on two feet also exacts a toll in wear and tear—a fact that medical specialists, who are experimenting with new treatments for everything from leg cramps to sprained ankles, and America's 34 million joggers know all too well.

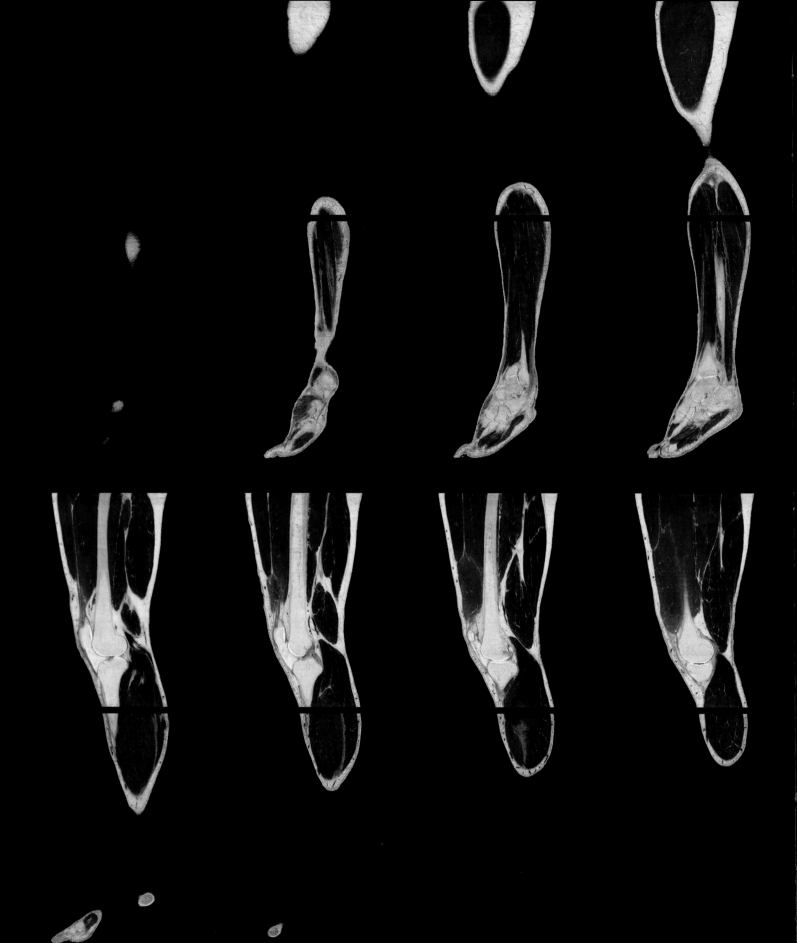

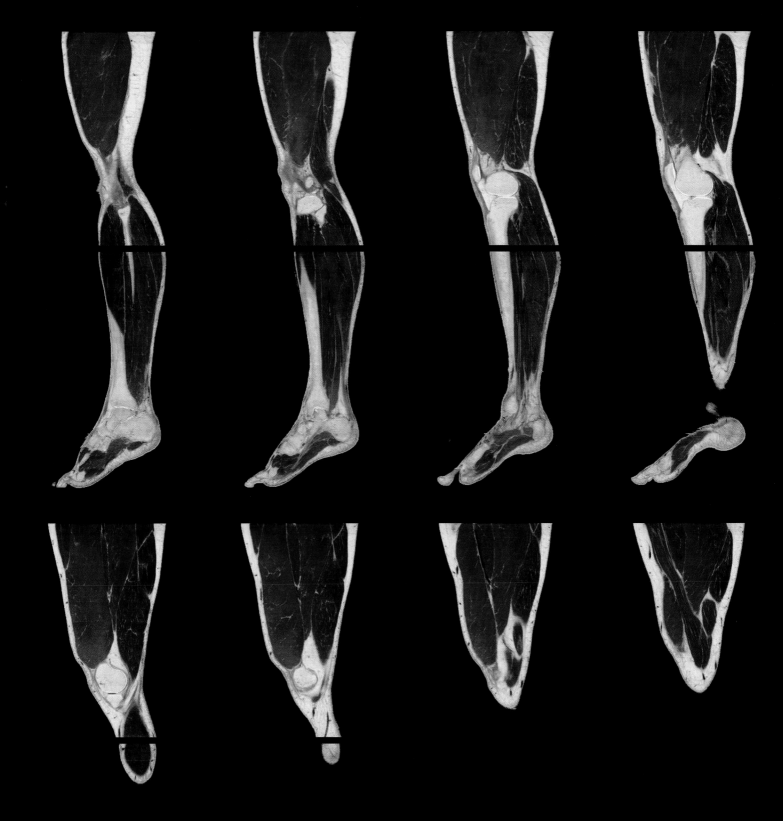

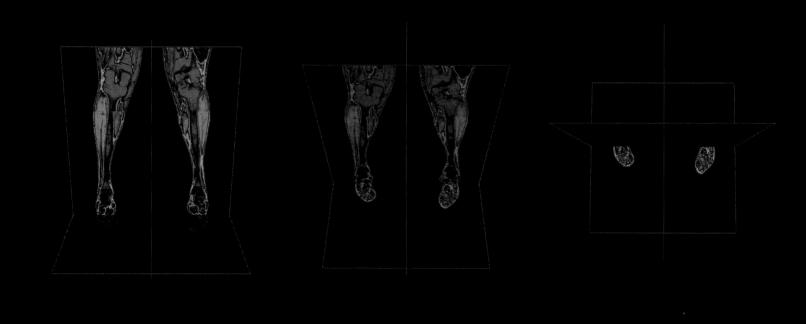
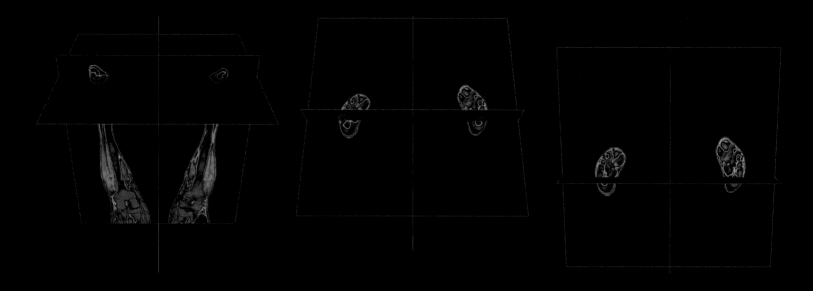

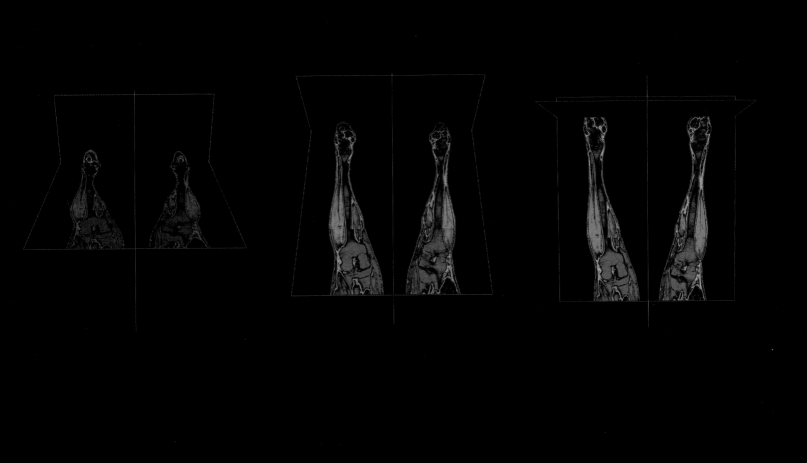
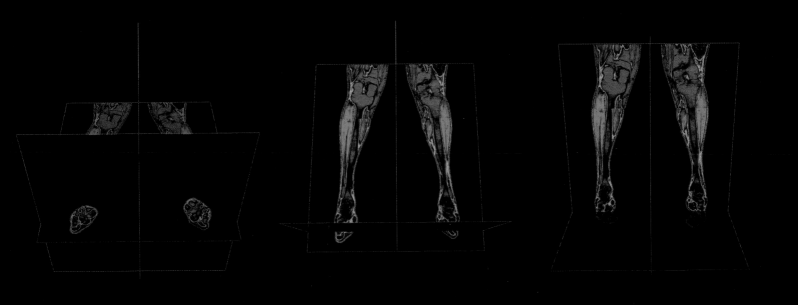

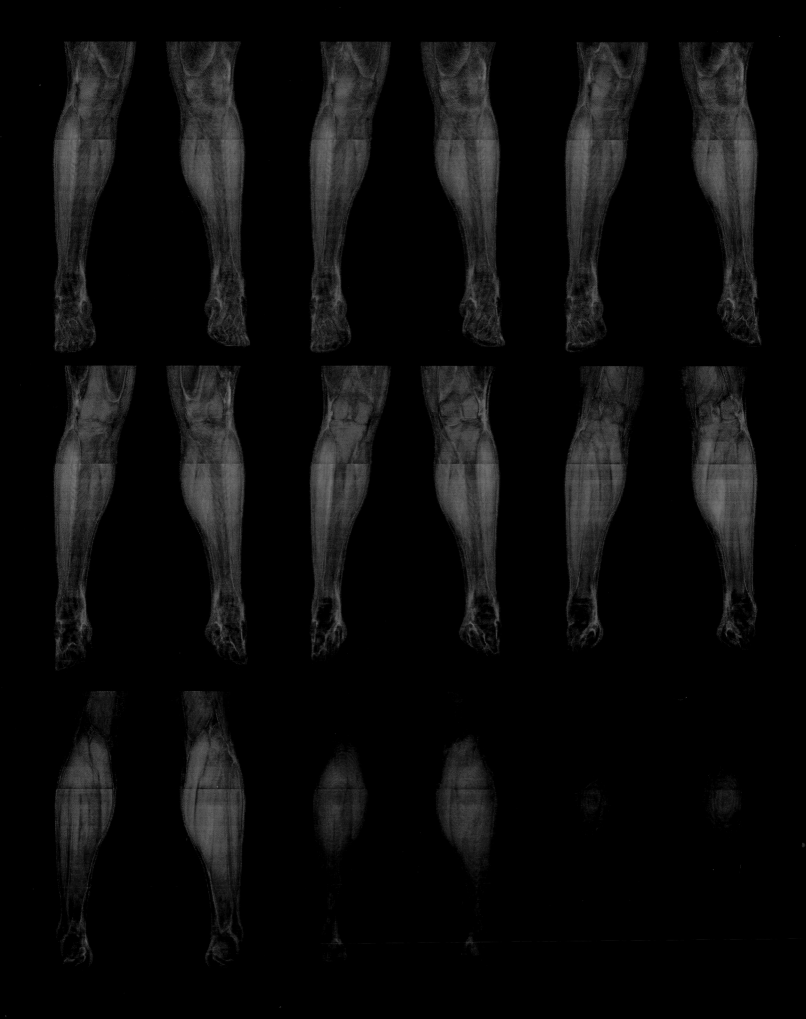

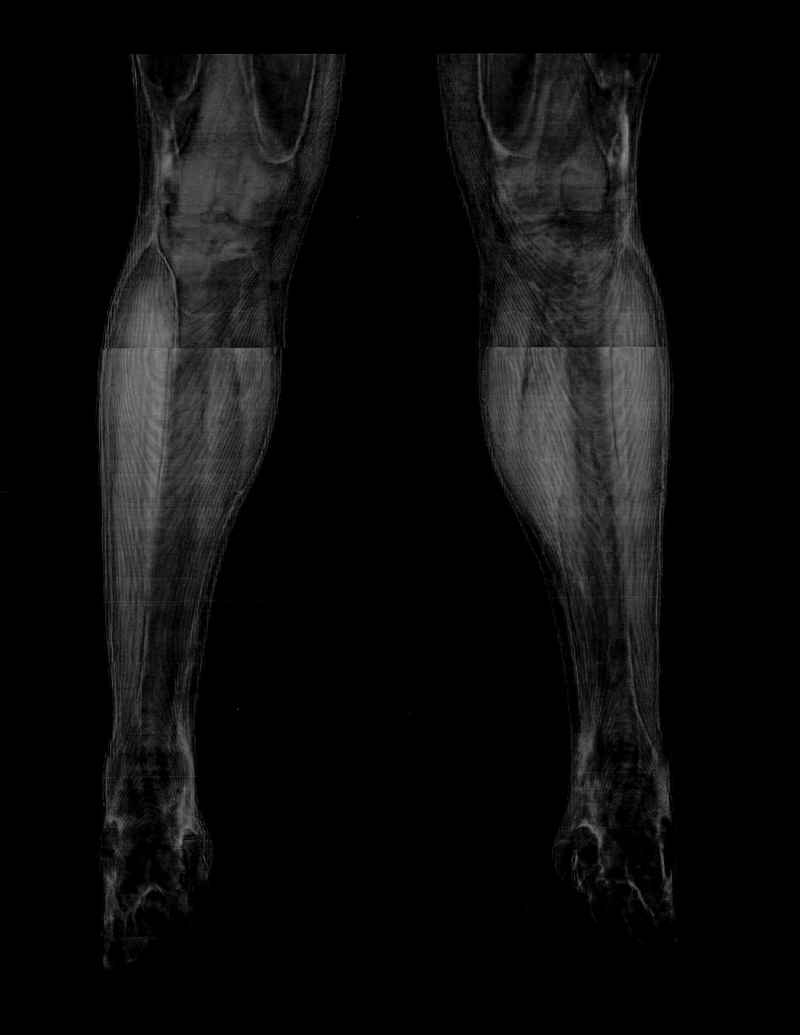

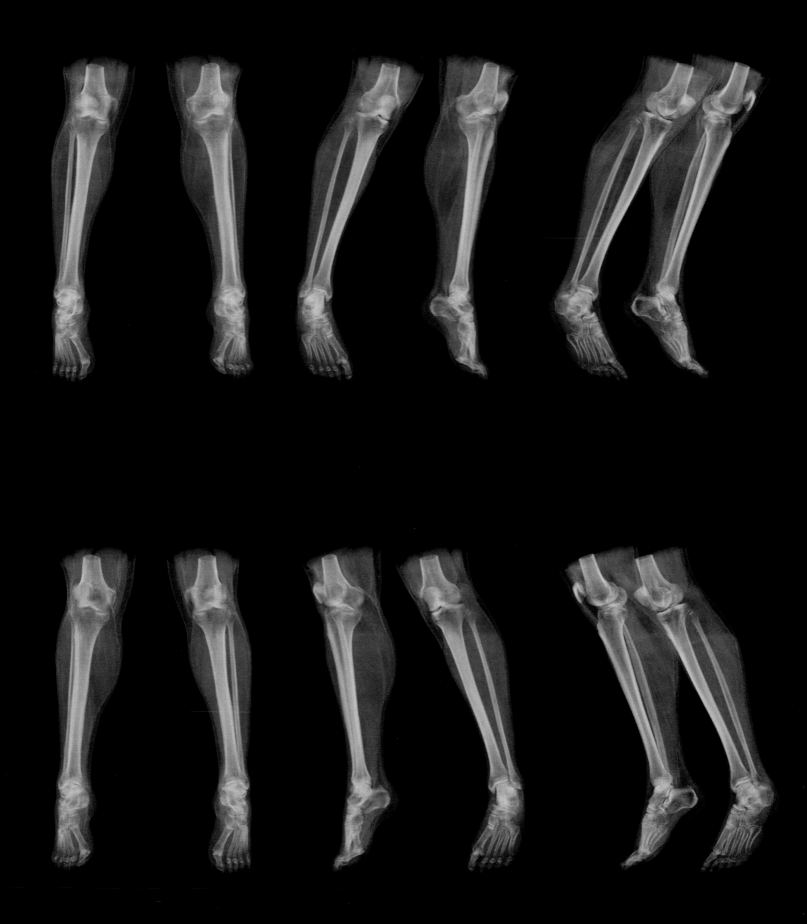

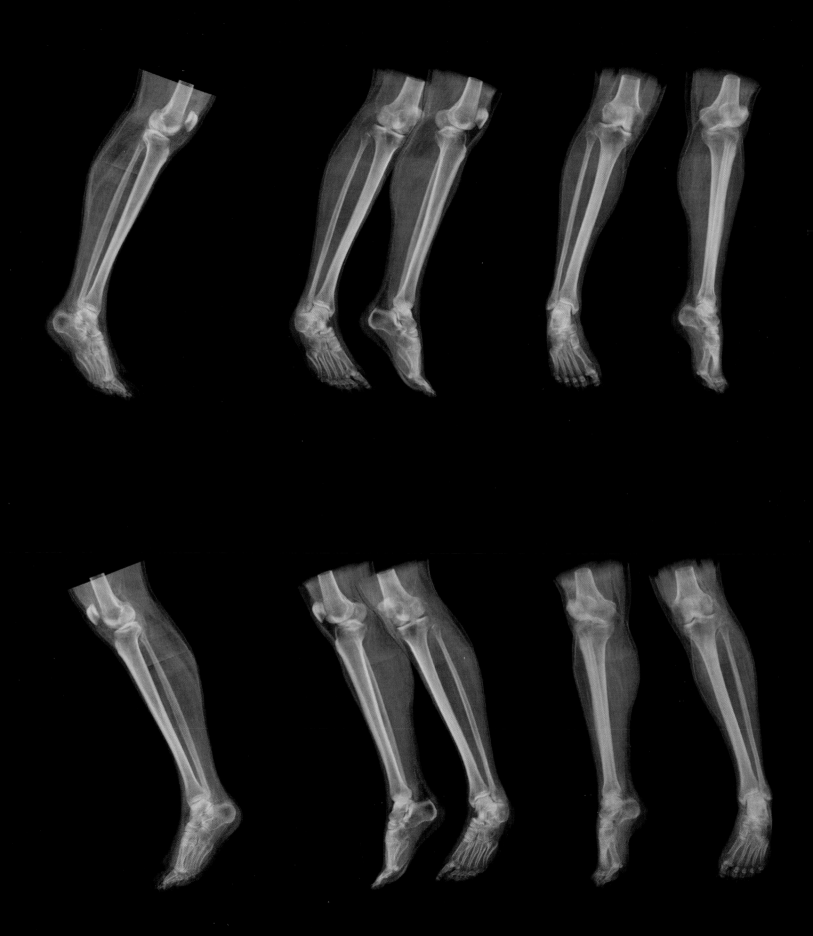

A QUICKER FIX FOR SPRAINS:

Last year, Temple University volleyball star Christine McGeough underwent an unusual form of therapy for a severely sprained ankle. She was sealed tightly in a pressurized oxygen tank, similar to the hyperbaric chambers used to treat divers suffering from the bends. During a 90-minute session, the injured athlete breathed pure oxygen as the air inside the chamber reached twice normal pressure. Two weeks and two sessions later, McGeough, who would normally have been sidelined for at least a month, was back on the court.

Time—along with a regimen of RICE (rest, ice, compression in a tight wrapping, and elevation)—remains the recommended therapy for sprained ankles, the most common form of all joint injuries. Ankles are especially vulnerable to the tears and strains called sprains. These complex hinges of bone and muscle, ligaments and tendons, support your entire body and, when you run or jump, may transmit a force of impact equal to three times your weight.

The benefits of hyperbaric therapy, which is thought to reduce swelling and hasten recovery by increasing the flow of oxygen to injured areas, were first demonstrated in 1993, when a study of Scottish rugby players showed that sprains healed 67 percent more quickly after pressurized oxygen treatments. Since then, the Vancouver Canucks, Dallas Cowboys, and San Francisco 49ers have invested in hyperbaric chambers for their players. However, the costly treatment is still not widely available outside of professional sports.

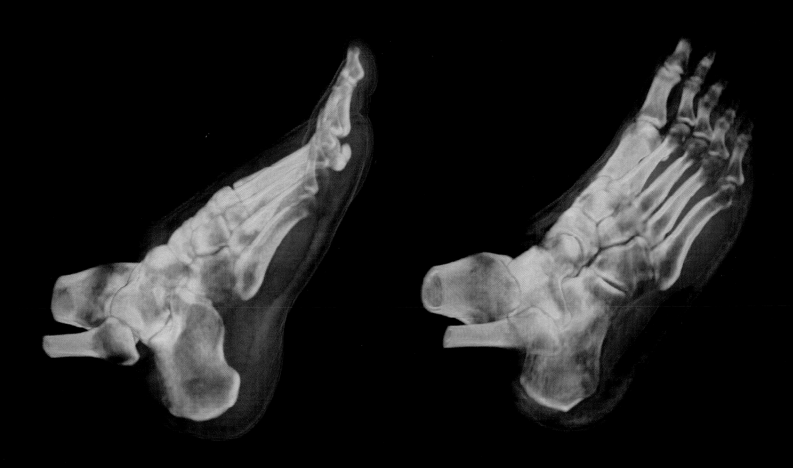

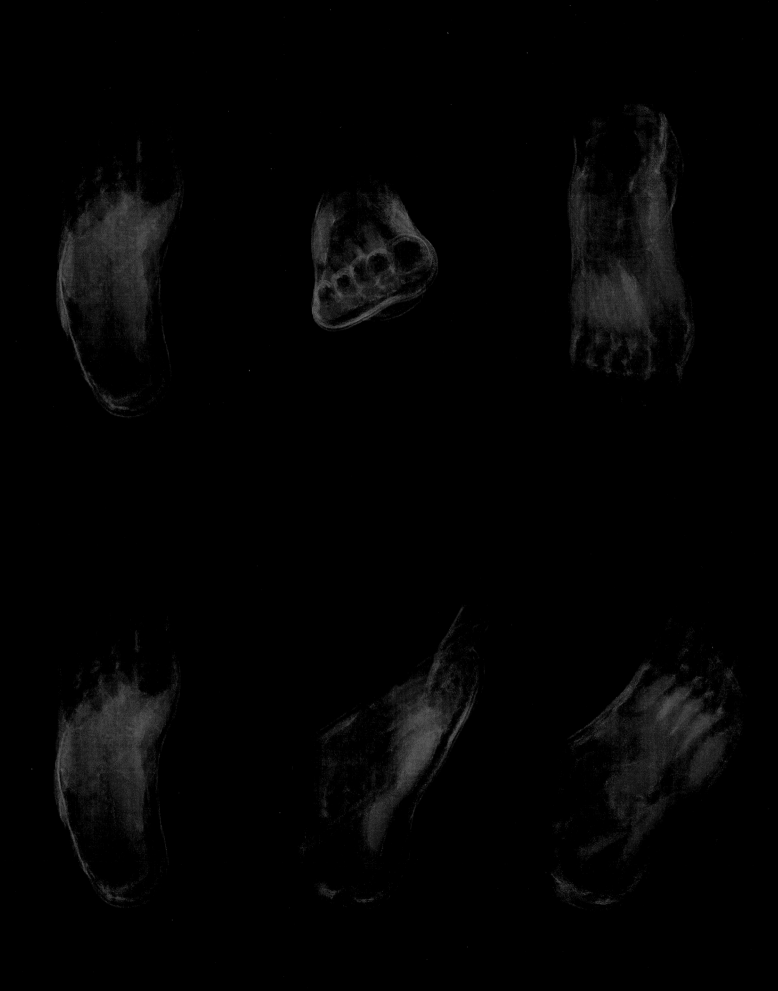

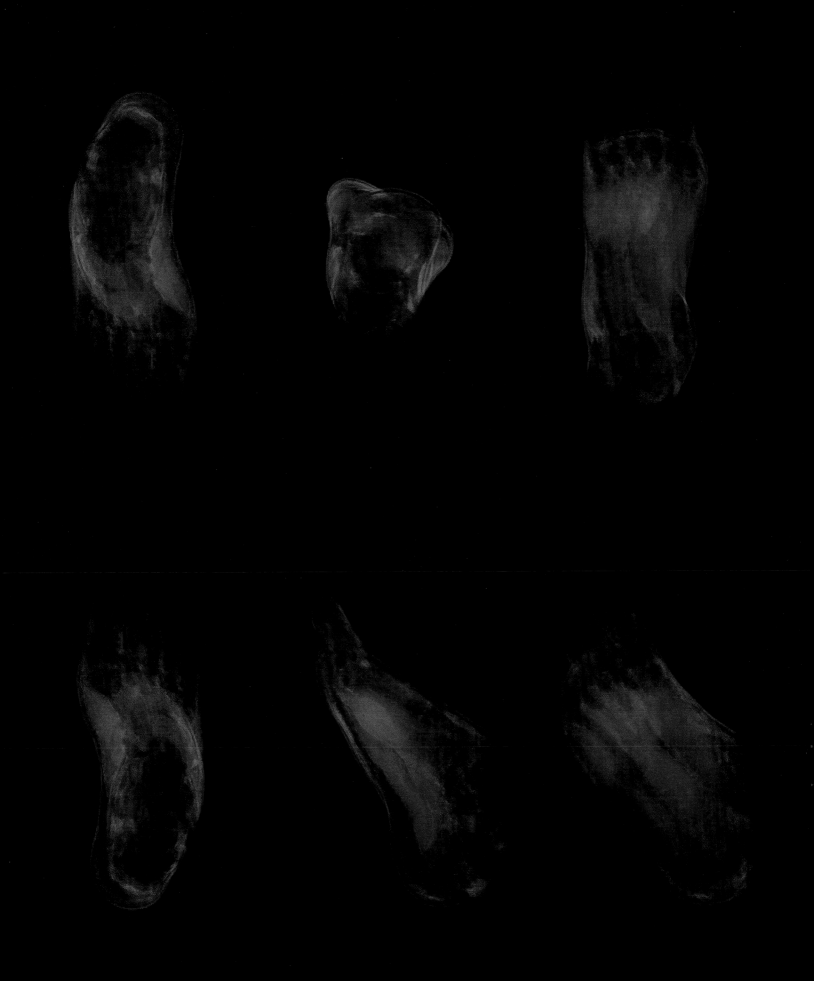

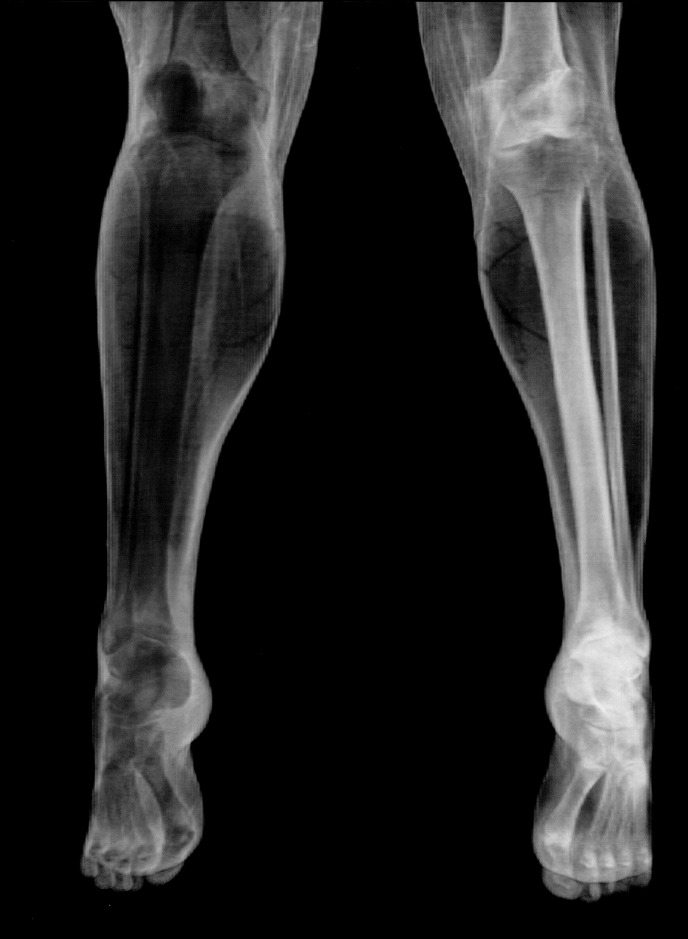

PHANTOM LIMBS:

The patient's leg had been amputated above the knee many years earlier. Yet his leg was still unmistakably, sometimes painfully present. "There's this thing, this ghost-foot, which sometimes hurts like hell—all the toes curl up or go into spasm," he told Oliver Sacks, who described the case in his book *The Man Who Mistook His Wife for a Hat*.

Patients who lose a leg or arm as a result of surgery or accident often report the feeling of a "phantom" limb. They can plainly see that the physical limb is no longer there. Nevertheless, they feel an image of it. The phantom may flare up in pain or seem to move its toes or grip things. Over time, it may change shape in peculiar ways. For instance, a phantom leg may disappear, to be replaced by a giant, leg-sized foot, or a phantom arm may dwindle into a wiggling hand attached to the shoulder.

Phantom limbs can be a painful, maddening ordeal. Surgeons have tried, without success, to exorcise them by cauterizing nerves leading from the amputation site. Researchers now believe that the sensation may actually arise in the brain itself, which overcompensates for the missing limb by expanding rather than reducing the number of sensory neurons assigned to the amputation site.

Yet phantoms do perform one beneficial function. Amputees who can learn to incorporate their phantoms into artificial limbs use their prostheses more successfully. Putting a phantom to work can also affect the way it feels. When Sacks's patient put on his artificial leg and walked, his phantom pain disappeared. "The leg felt good," he reported. "It animates the prosthesis and allows me to walk."

THE CASE OF THE MISSING TOES:

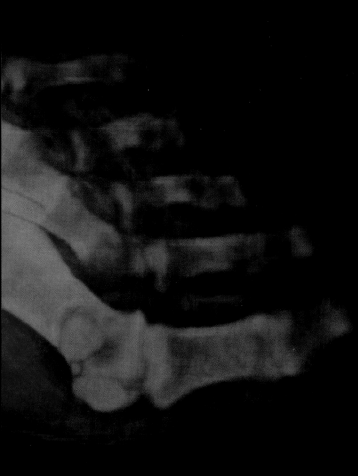

In 1988, paleontologists working in Greenland unearthed a set of fossilized bones belonging to a giant salamander-like creature called Ichthyostega, which lived over 500 million years ago. The find contradicted a belief held by evolutionists since the days of Darwin that, when our earliest ancestor crawled up on land, it did so on limbs equipped with five digits each. The pentadactyl ("five-fingered") limb was, in fact, considered to be a prototype for humans and all other "four-legged" tetrapods. But the Greenland bones told a different story. Ichthyostega, one of the oldest tetrapods, had seven toes on its back legs and eight on its forelimbs. The discovery suggests that our own five-digit feet and hands evolved as nature gradually reduced the number of digits by natural selection.

Experts have argued that the pentadactyl limb was an ideal configuration for bearing weight while walking, providing just enough support to prevent wobbling without being so cumbersome that it interfered with locomotion. But the design was particularly fortuitous for humans, since it provided both a five-toed foot to walk on and a five-fingered hand, vital for the use of tools, not to mention the development of our base 10 system of arithmetic. Yet, the case of the missing toes is far from solved, according to the naturalist and author Stephen Jay Gould. If five toes are needed to bear the weight of the body, why, Gould wonders, did nature construct the human big toe as the foot's main weight-bearer? And why was the numbers of digits further reduced to four in the hooves of horses, cows, deer, and other surefooted mammals? The pentadactyl limb may be one of the great achievements of human evolution, but according to Gould, our glorified five digits may actually have arisen out of chance, not necessity.

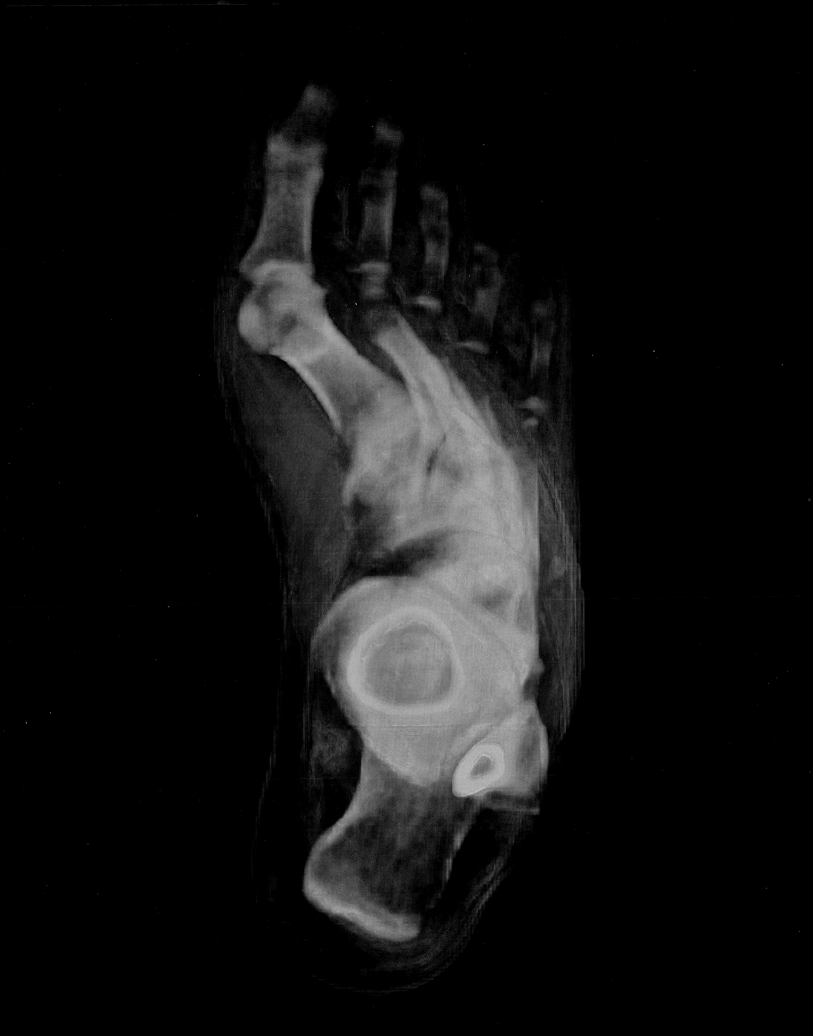

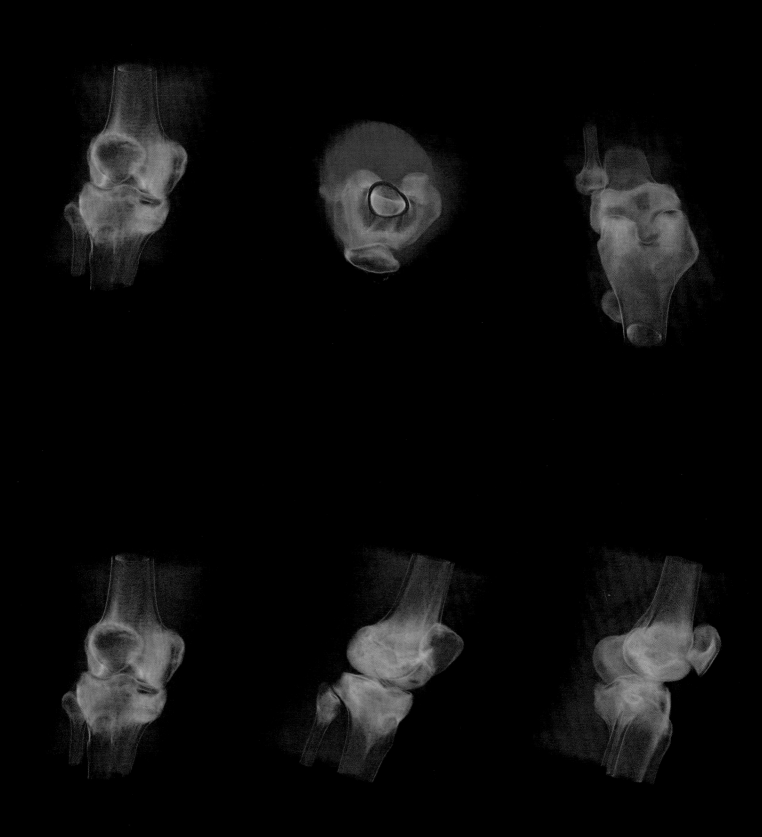

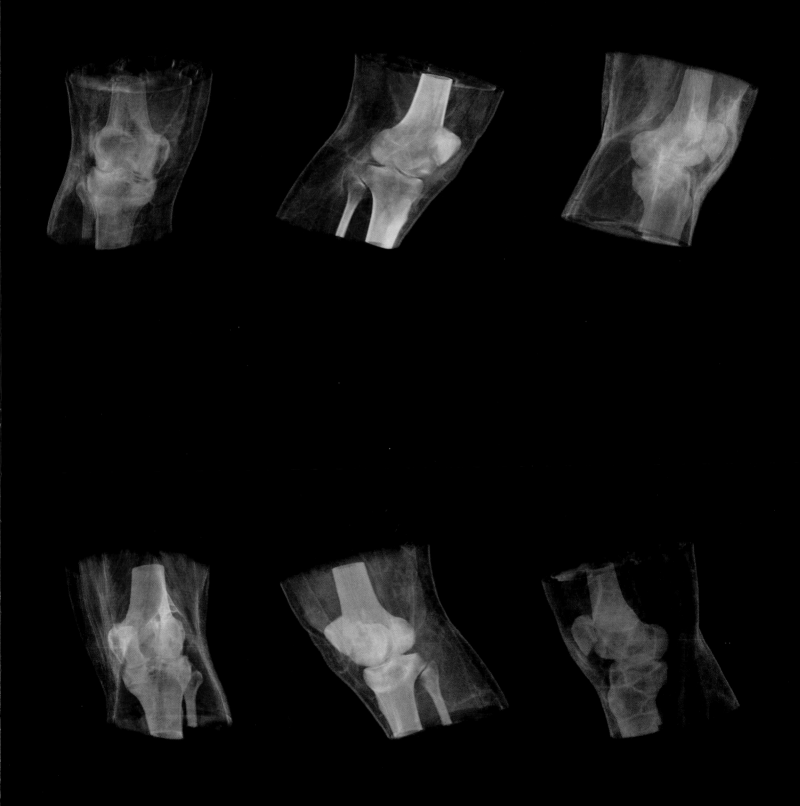

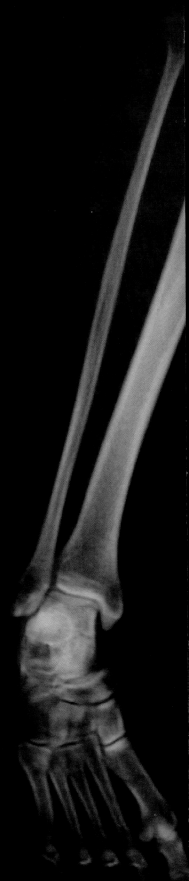

FIRST STEPS:

The discovery of the "Lucy" skeleton in Hadar, Ethiopia, in 1974 and the 1995 finding of an even older set of bones near Kenya's Lake Turkana have provided anthropologists with graphic evidence that a new type of creature, walking upright, was roaming the earth some 4 million years ago. While these ancient hominids retained many apelike features—a small brain, small ears, and large canine teeth—their leg bones were distinctly human. The tibia (the main shin bone) had grown thicker at both ends to bear the weight and the stress of upright walking, and a knob on top of the tibia was trumpet-shaped, rather than T-shaped as it is in chimpanzees, creating the more stable human knee joint needed for balance.

Scientists believe the transformation from ape to modern Homo sapiens took over 5 million years to complete. During that time, to facilitate walking upright, large new buttock muscles were added, the knee acquired a locking mechanism, and our feet were completely remodeled—from the handlike appendages that enable apes to grasp branches and objects with their toes, to the human design, in which the big toe stabilizes our stride and gait.

Why the earliest humans stood upright remains an unsolved mystery. Evolutionists agree that walking upright is the most unstable of all modes of locomotion and would never have been resorted to as an "improved" means of getting around. Instead, scientists have proposed several alternative theories. Standing upright may have helped Lucy and her kin to spot lurking predators in the prehistoric savannah, lessened the heat our ancestors absorbed from the tropical sun, or merely enabled them to keep their heads above water during times of flooding.

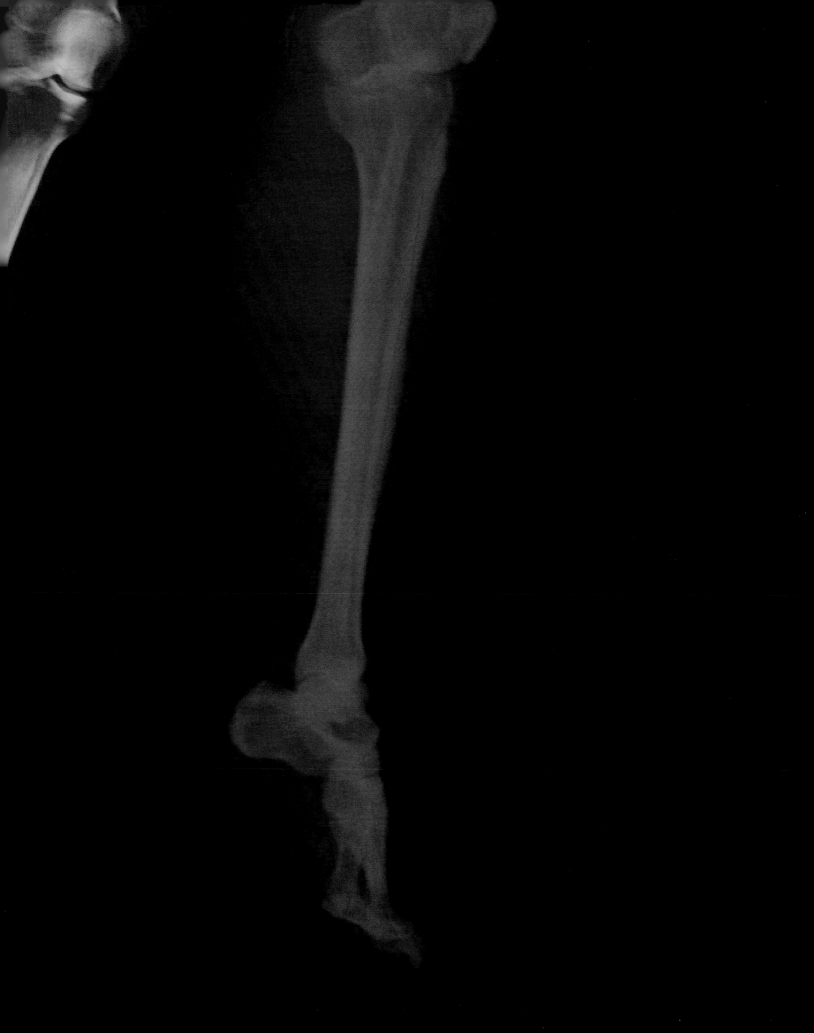

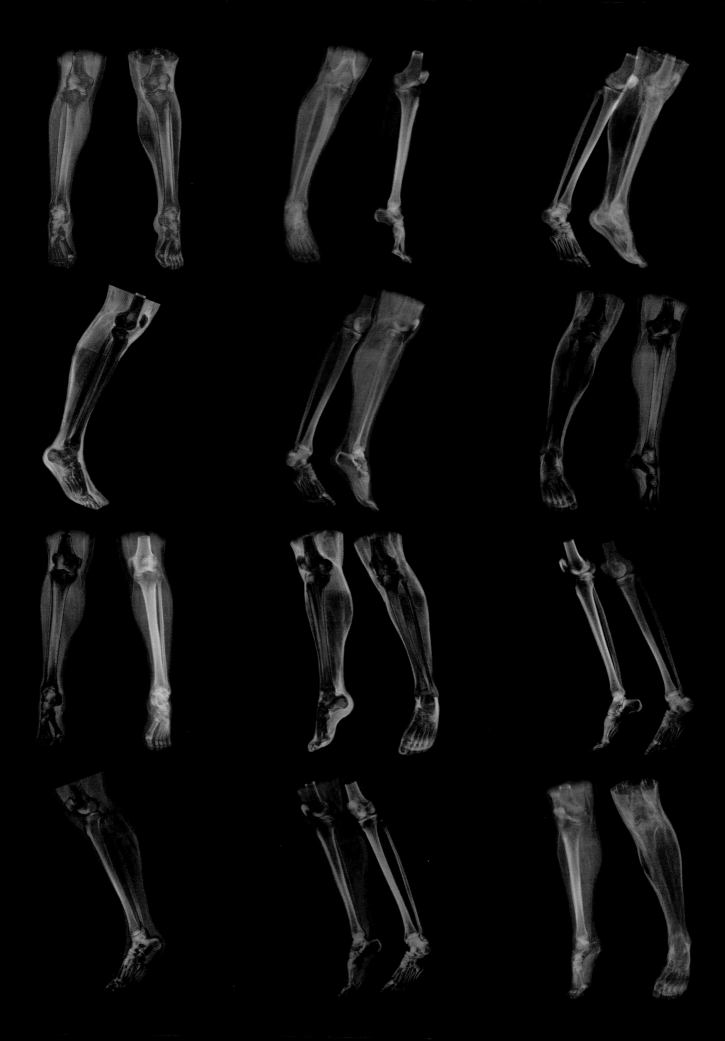

WALK, DON'T RUN:

Aerobics, the 1968 best-seller by Dallas physician Kenneth Cooper, revolutionized Americans' attitudes about physical fitness. Before its release, only 100,000 Americans called themselves joggers; today, more than 34 million people run regularly. Cooper's assertion that "vigorous" activity could prevent heart disease and other illnesses was echoed last year when a 26-year, 17,321-subject study by researchers at Harvard and Stanford Universities showed that strenuous forms of exercise like jogging can lower the risk of death by 25 percent.
However, Cooper himself now disagrees with the "no pain, no gain" premise. In fact, his own 8-year, 13,334-subject study revealed that even a "moderate" amount of exercise—a brisk half-hour walk once a day—is enough to cut the risk of dying from cardiovascular disease to half that of nonexercising couch potatoes.

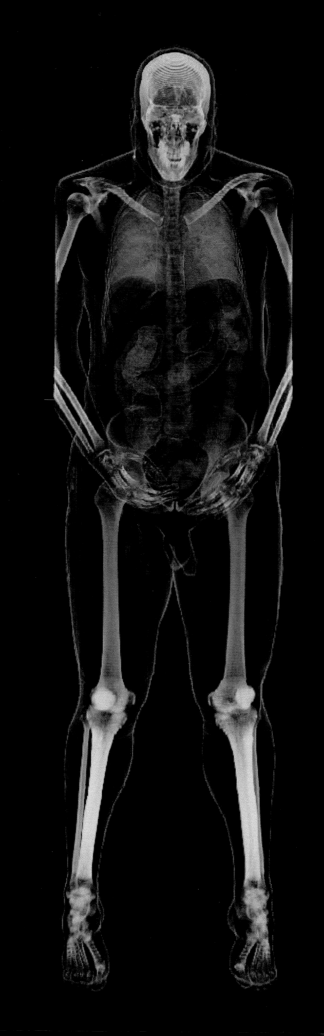

The Body

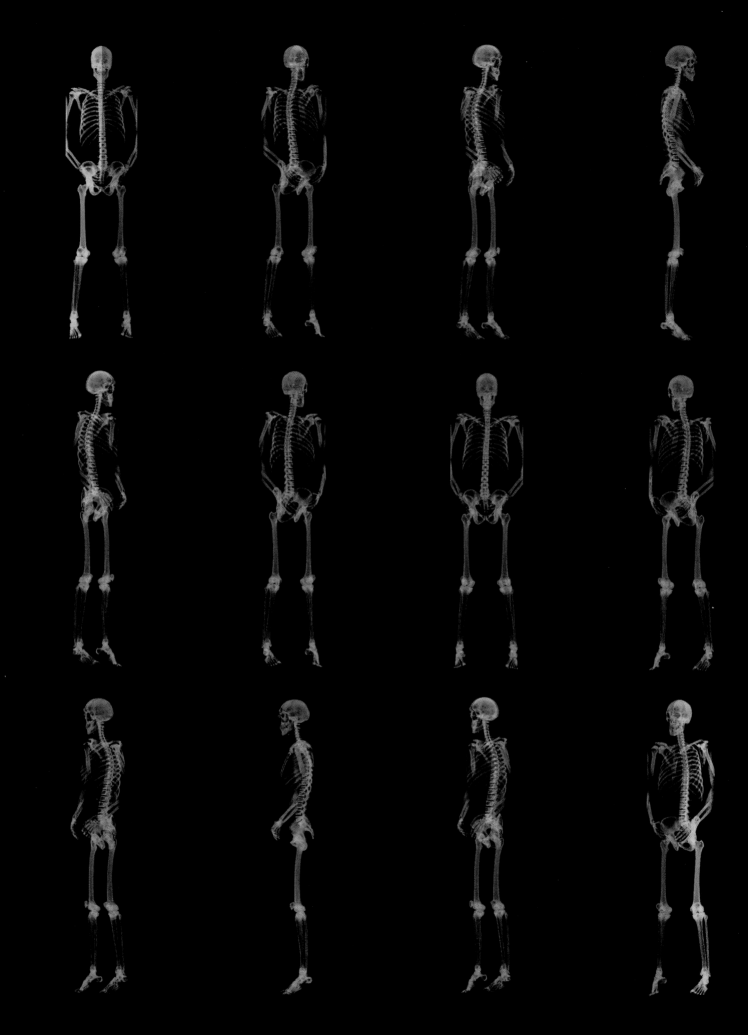

The anatomist and author F. Gonzalez-Crussi once wrote that the body's interior is "sacred space," which is never penetrated without "fear, awe, or passion." The digitized cadaver of Joseph Paul Jernigan provokes the same timeless fascination that has captivated artists, poets, philosophers, and scientists alike through the ages. Within these confines of flesh, blood, and bone are hidden the secrets of life itself, the "mysterious spark," as Gonzalez-Crussi wrote, "that propels life's flow and animates its stubborn pulsations."

The quest to unlock the secrets of the human anatomy is constantly generating new discoveries about the processes of life, from the formula for building human bone to the intricate mechanisms that make muscles move, and to the multifaceted functions of the body's largest organ, the skin.

Here, form and function merge into a living, breathing, thinking organism, a human being, a man—until disease and death take their inevitable toll.

Jernigan's anatomy tells us a *little* about him—the brawny muscles testify to years of weight-lifting in the prison gym; the layer of fat encasing his body to a diet of greasy jail food. His internal organs appear free of disease (the naked eye cannot detect the lethal dose of drugs with which he was executed). By enabling us to peer into the body's hallowed space, these images tell us *a lot* about ourselves

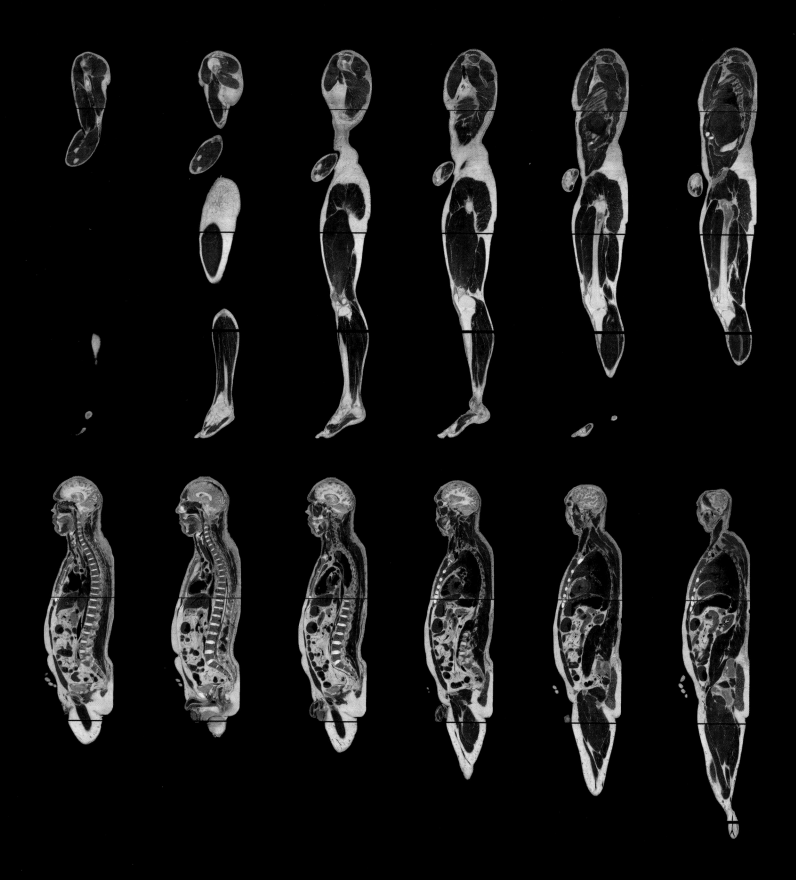

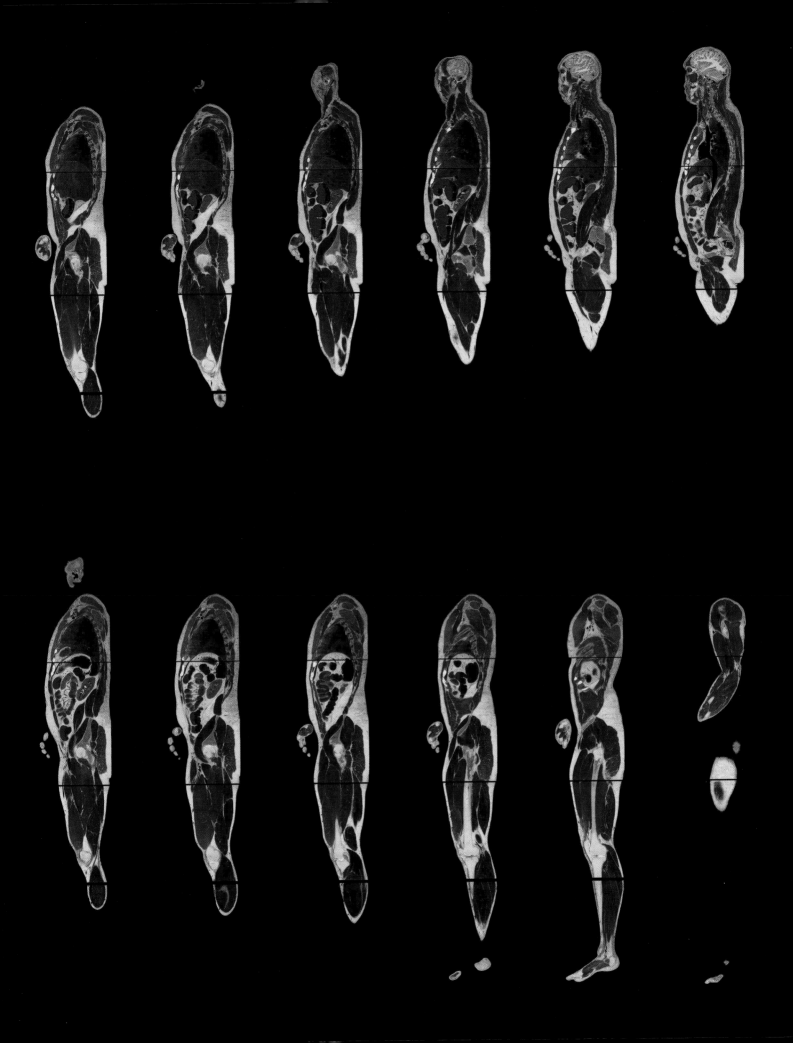

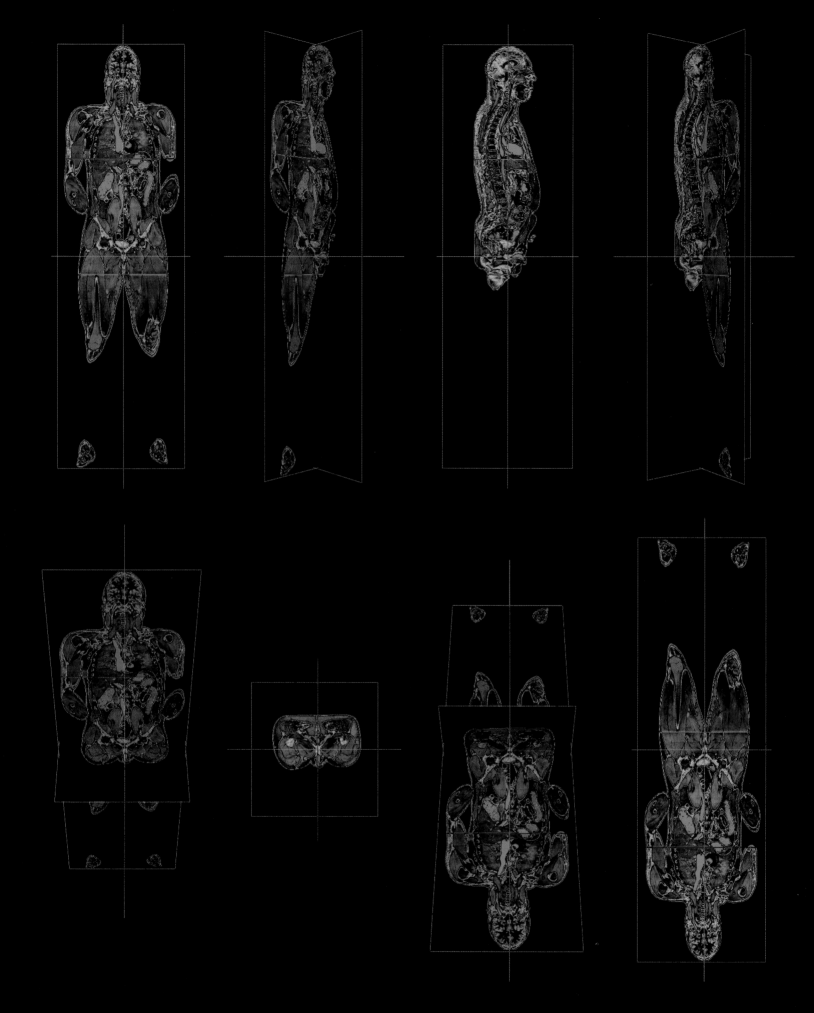

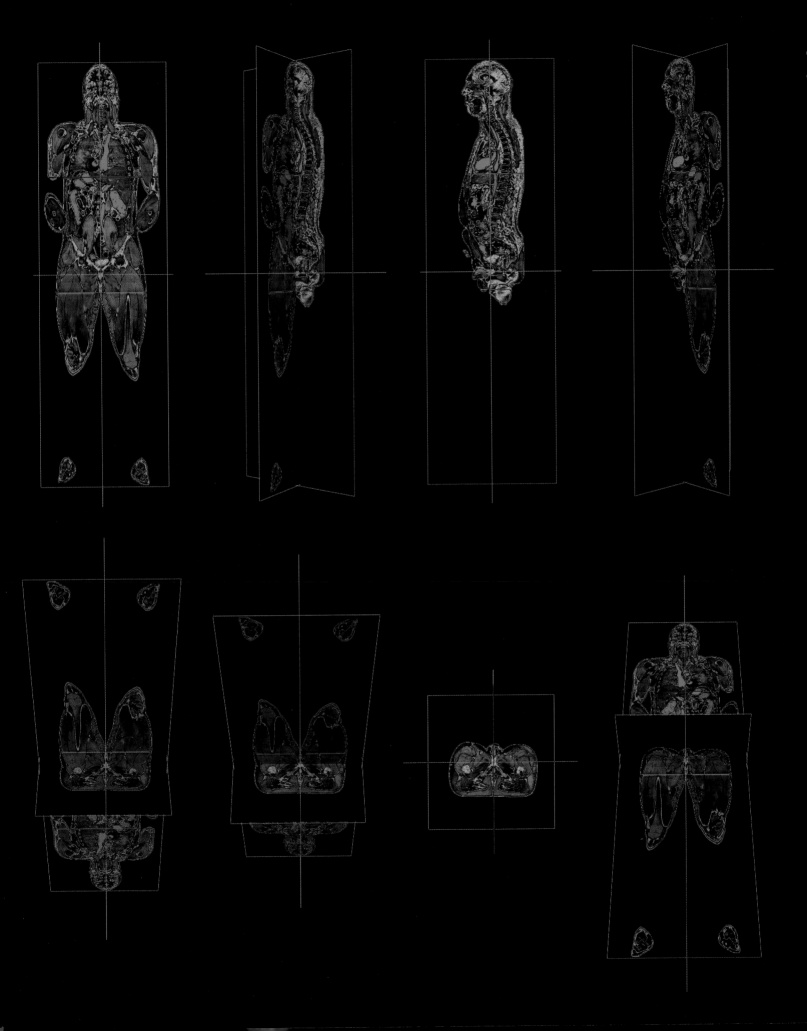

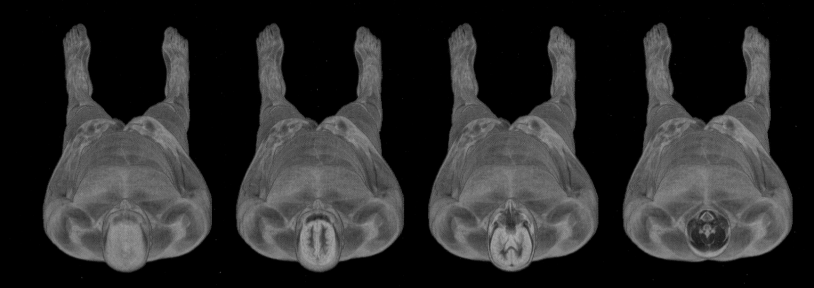
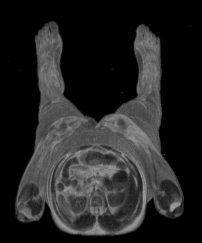
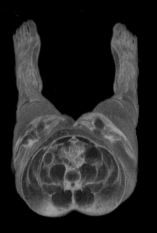
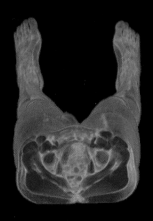
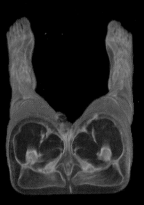

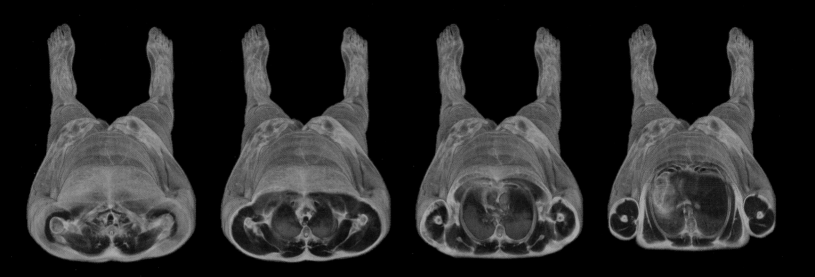

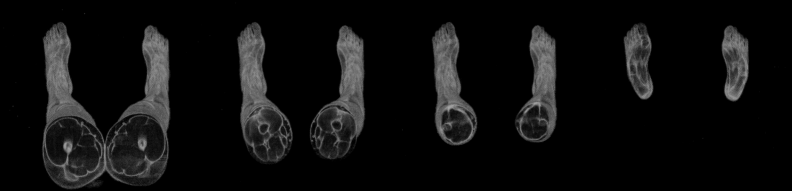

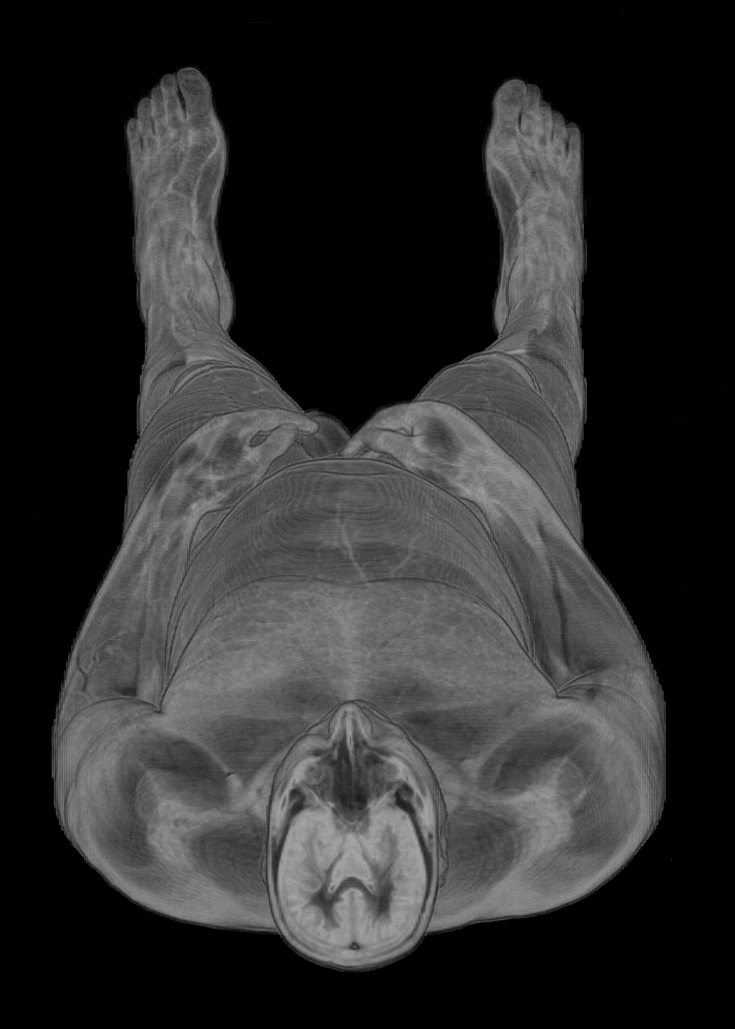

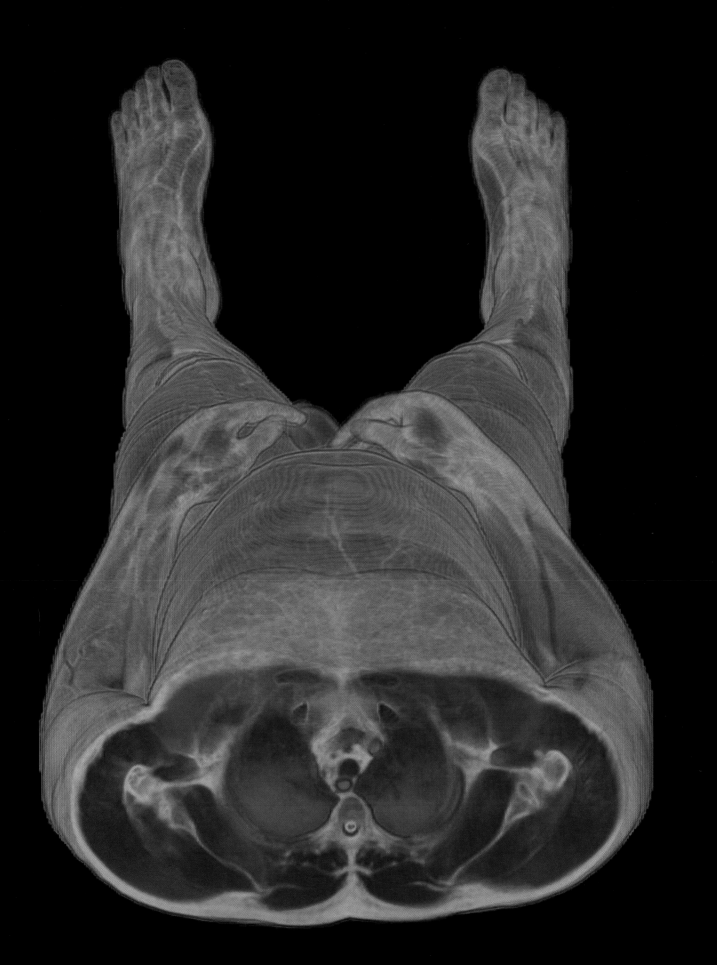

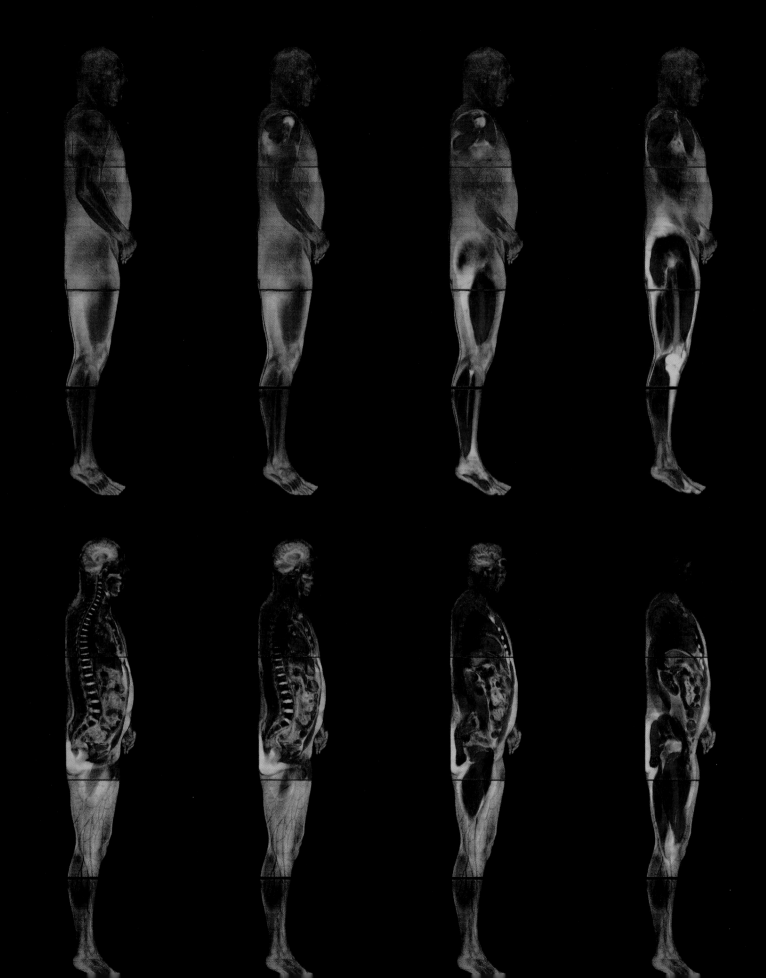

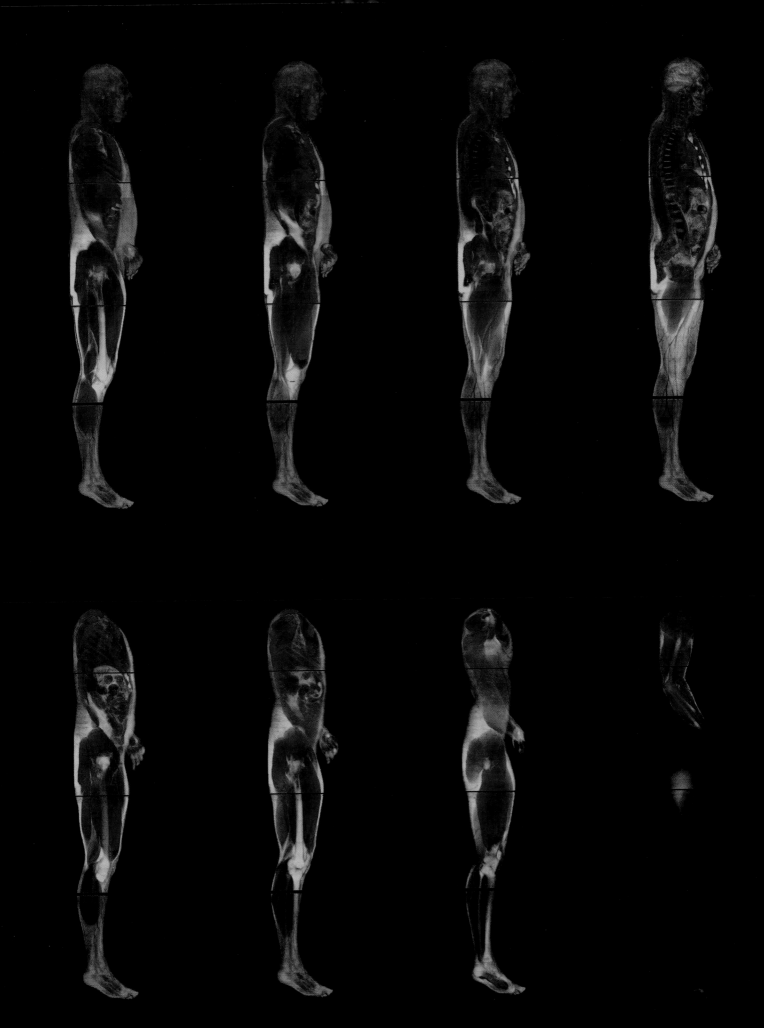

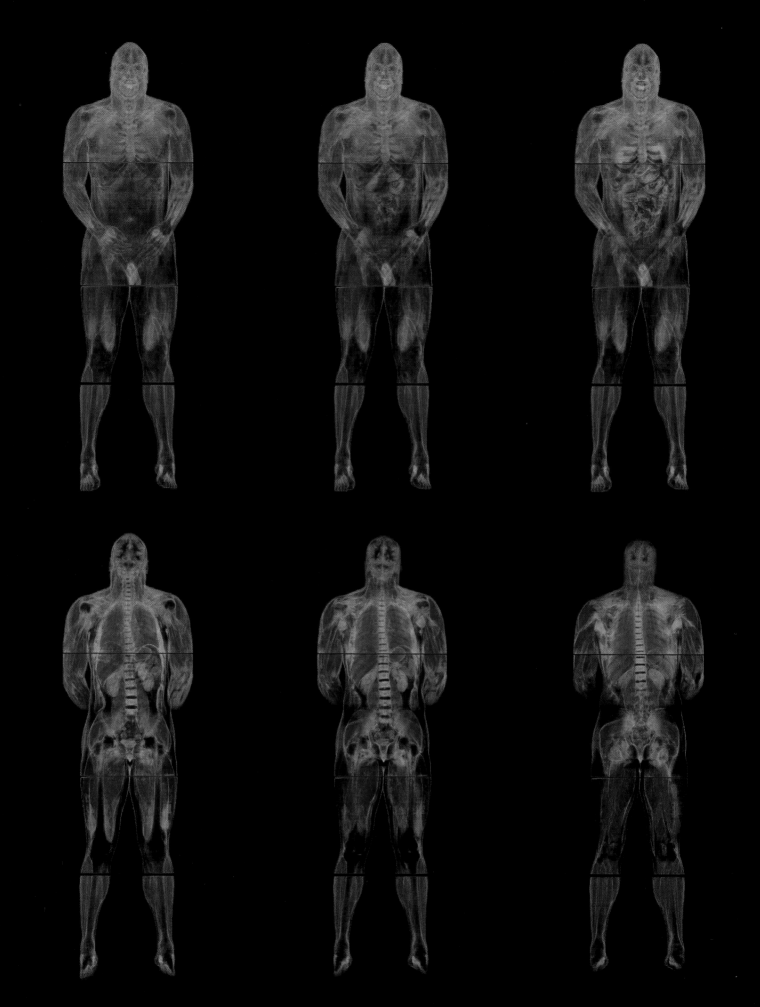

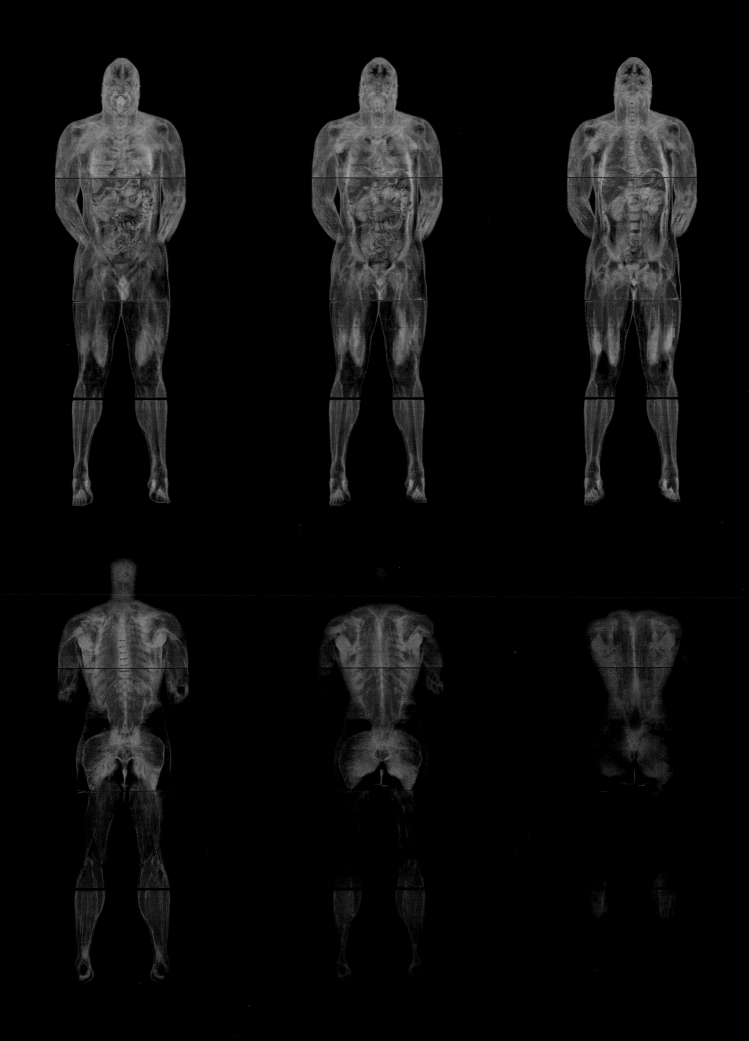

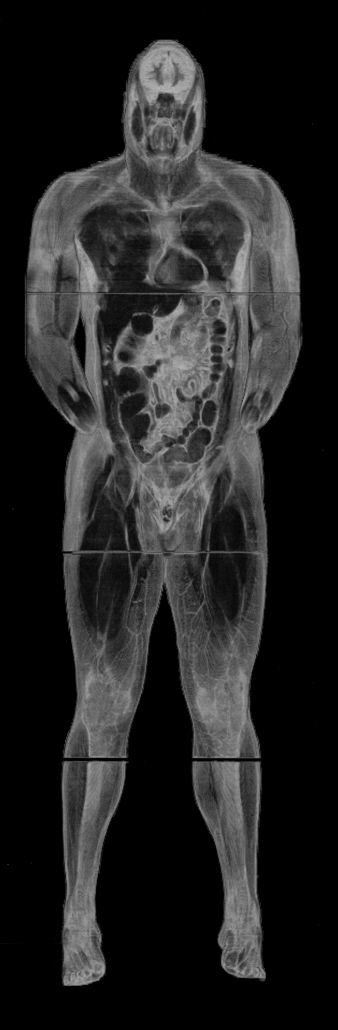

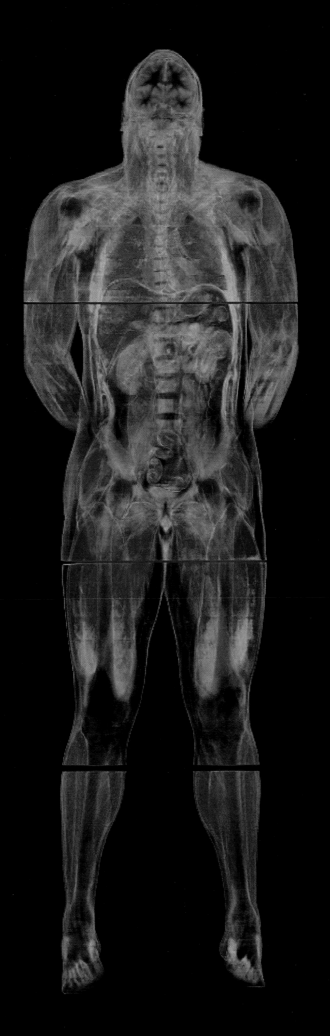

SKIN-DEEP SECRETS:

The British zoologist J. Z. Young once wrote that the main function of skin in living organisms is to "prevent their dissolution into the surroundings." However, scientists now recognize that skin is much more than a fleshy bag holding us together. Measuring roughly 2 yards in surface area and weighing between 6 and 9 pounds, the skin is not only the body's largest organ but also one of its most complex and versatile. A half-inch-square piece of skin contains more than 3 million cells nurtured by 3 feet of tiny blood vessels, in addition to some 80 sweat glands and 35 nerve endings, to help us keep cool and touch the world around us.

Nearly transparent and as thin as 2/1000 inch over the eyelids, the part of the skin we can see, its outer layer, or epidermis, is continually being replenished as cells at the surface die, are shed by the millions during every shower, and are then replaced. The dead cells leave behind a layer of hard protein called keratin—the same substance composing hair and fingernails—which coats the skin with a tough, protective crust to keep out dirt and germs. Beneath the surface, in the skin's interior layer, or dermis, sebaceous glands release a unique fatty substance called sebum, which lubricates and waterproofs the skin, so that it remains soft and pliant and doesn't swell like a sponge whenever we take a bath.

Sebum also performs a vital, final role in the body's immune defenses against disease, by flushing toxins and dead cells from the glands to the surface of the skin. Acne develops when hormonal changes during puberty cause the sebaceous glands to work overtime. Excess sebum blocking the ducts of the glands results in an accumulation of dead cells and melanin—the pigment that gives skin its color. The result: a darkened plug, or blackhead.

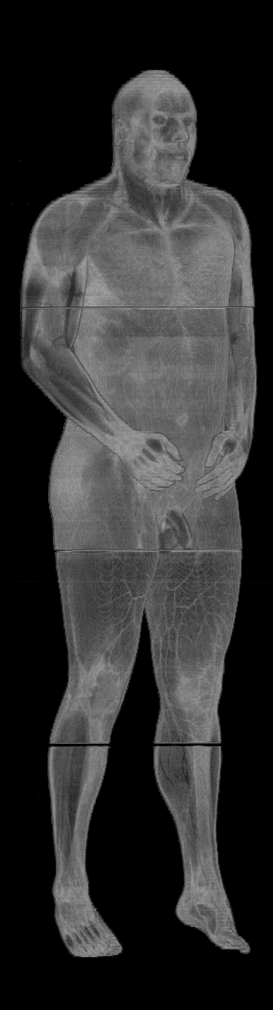

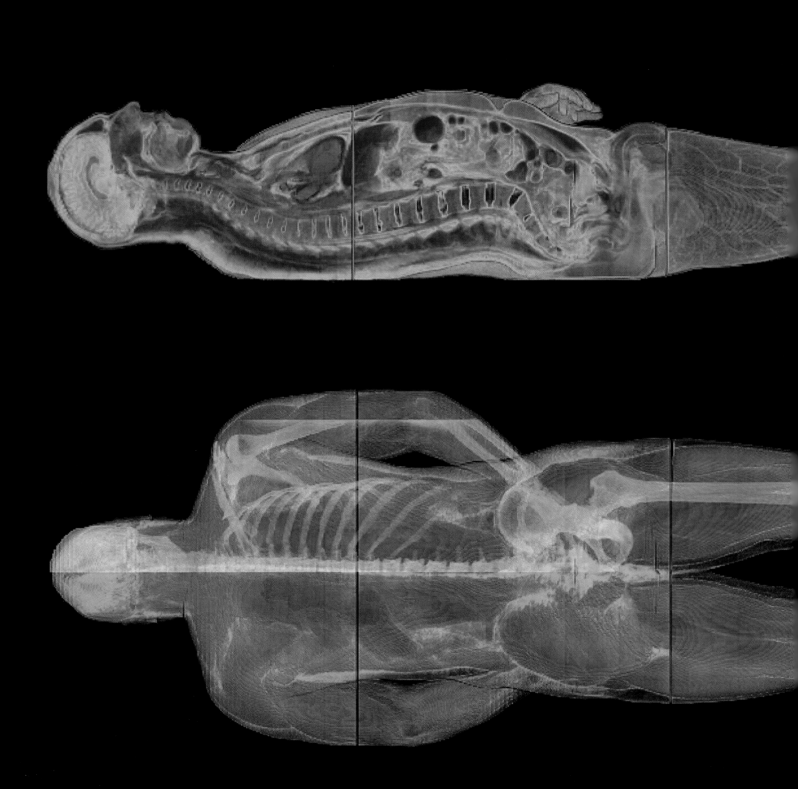

IMPROVING THE ODDS FOR SPINAL CORD INJURY:

When a riding accident shattered his first two vertebrae, rupturing his spinal cord, actor Christopher Reeve's chances of survival—much less of recovery—were slim. Every year, spinal cord injuries are sustained by over 12,000 people, mostly young men involved in automobile accidents, with sports-related accidents accounting for an additional 14 percent; just over half live.

Because the millions of densely packed nerve cells in the spinal cord transmit the rapid flow of signals from the brain to the body's periphery, which enable us to move, sense the world around us, and function normally, spinal cord injuries can be catastrophic. Communication between the brain and the parts of the body below the injury is interrupted, and those areas lose sensation and the ability to move. When the injury is close to the top of the spinal cord, as it was in Reeve's case, the damage is the most disastrous. High spinal cord injuries can cause paralysis not only of the arms and legs, but also of muscles that control the chest and the diaphragm of patients, requiring them to use a respirator to breathe.

There are no cures for spinal cord injury—tissues of the central nervous system cannot regenerate—but improved trauma care and rehabilitation techniques have enabled patients like Reeve to make remarkable progress. The anti-inflammatory drug methylprednisolone, which Reeve received within hours of his injury, has also been credited with dramatically boosting the chances of at least partial recovery for spinal cord injury sufferers, by reducing swelling and increasing the flow of blood to surviving spinal cord nerves.

Physicians consider any improvement in motor and sensory function following spinal cord injury to be highly significant. In Reeve's case, the prognosis is promising. At first only able to flex the trapezius muscles in his back, the actor is now able to drive his electric wheelchair by "sipping" and "puffing" air through a plastic tube he holds in his mouth. He has also learned to speak audibly by timing his words to coincide with the exhalations of his respirator.

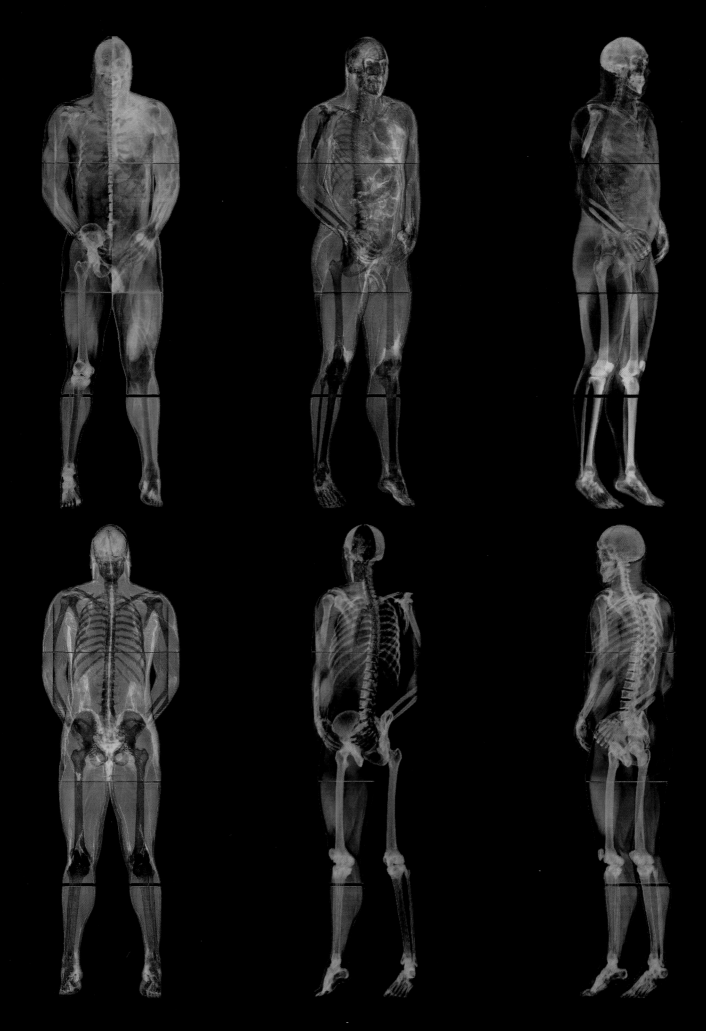

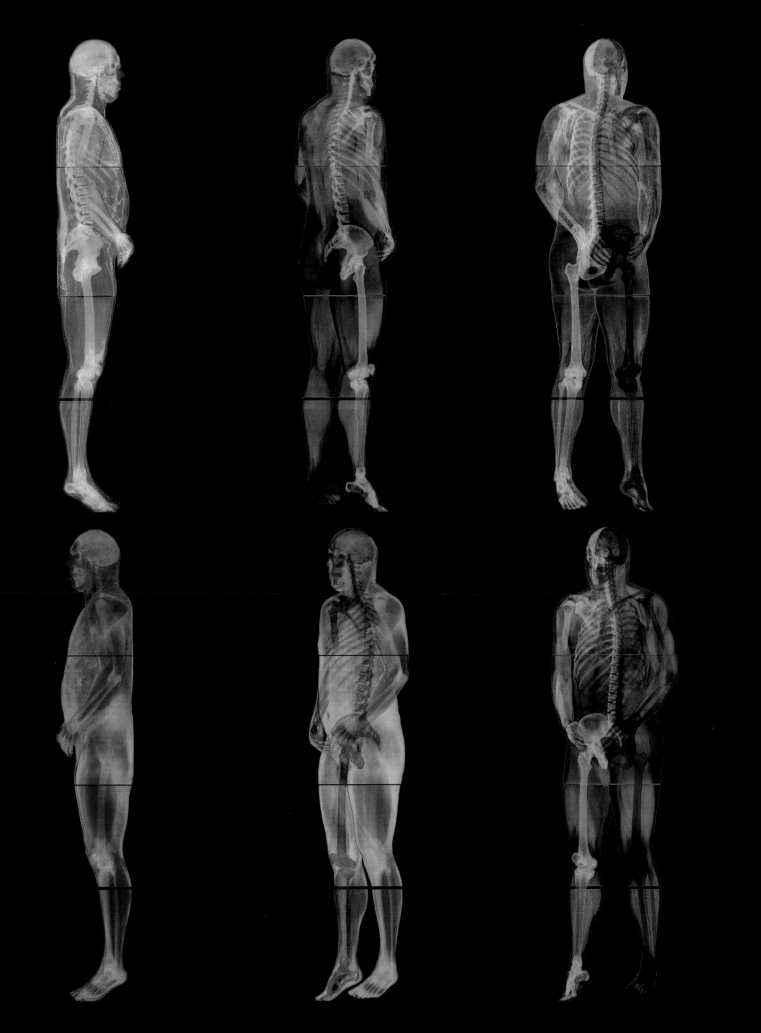

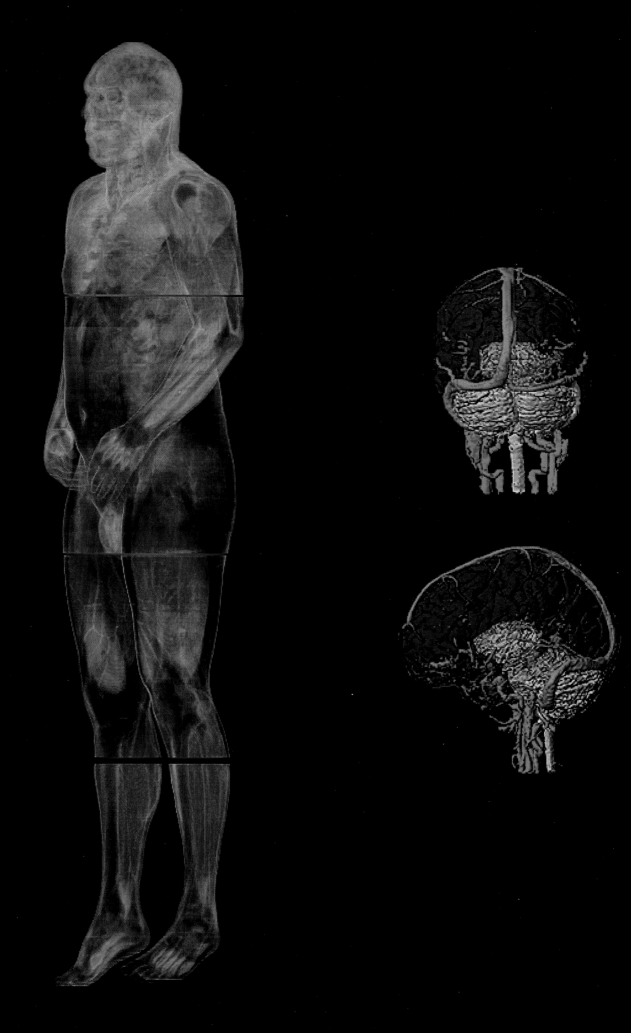

COOLING IT:

The human body is far from the world's most energy-efficient machine. Indeed, only about one-quarter of the body's energy is actually converted into work, while the other three-quarters is converted into heat, which is dissipated through radiation, breathing, and the evaporation of perspiration. At peak output, the 3 million sweat glands in the skin can produce up to a gallon of perspiration an hour to cool an overheating body.

This heat-regulating function is performed only by one type of sweat gland—eccrine sweat glands, which are located all over the body but are concentrated in the armpits, the soles of the feet, and the palms. Eccrine glands release an odorless form of perspiration through ducts opening directly to the skin's surface.

Less clear-cut is the function of apocrine sweat glands, which develop at puberty and open into hair follicles in the armpits, groin, and around the navel. Apocrine sweat glands produce "nervous" perspiration in response to emotional stimuli. The distinctive smell of body odor is produced when colonies of bacteria, feeding on organic compounds in apocrine sweat, produce a pungent chemical waste product—3-methyl-2-hexanoic acid. George Preti, a researcher at the Monell Chemical Senses Center in Philadelphia who isolated the compound, compared a flask containing a few drops of the chemical to "having an armpit in a jar." While apocrine sweat glands may have provided vital chemical mating cues for early humans, today they are mainly the target for the countless antiperspirants and deodorants, which work only on apocrine glands.

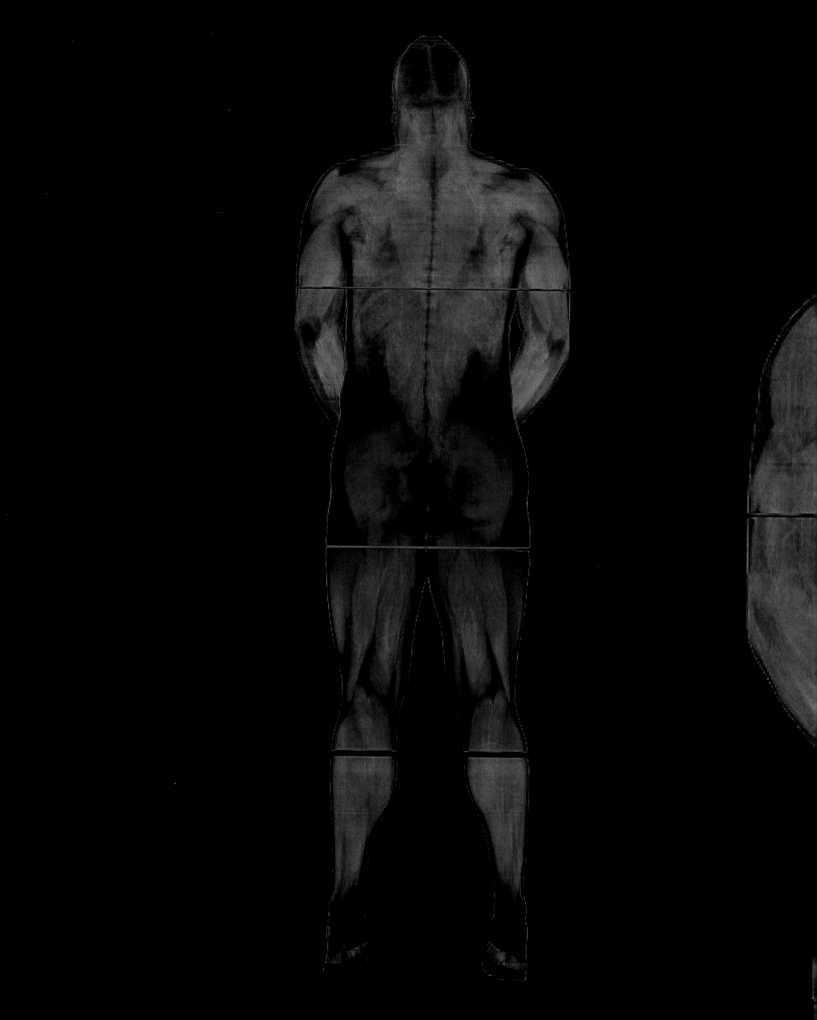

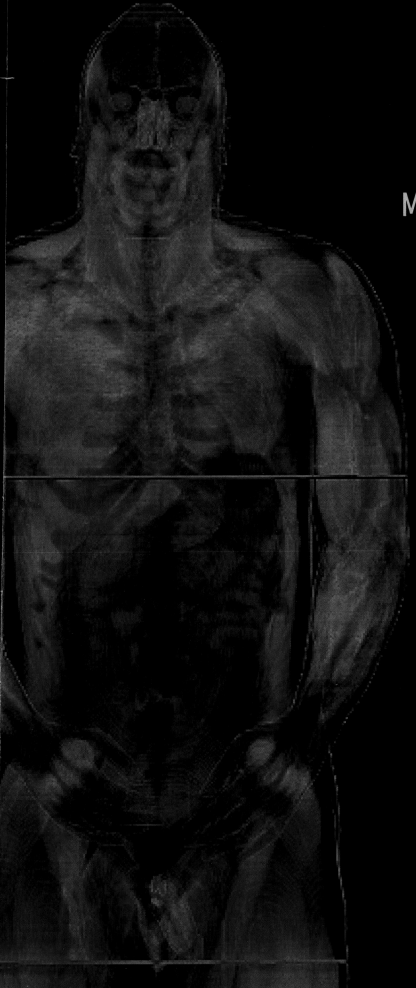

MUSCLES' MATTER:

Joseph Paul Jernigan's hulking physique was the product of years of lifting weights in the prison's gym. But while pumping iron can expand and sculpt skeletal muscle into Arnold Schwarzenegger's 22-inch arms or enable Olympic weight lifters to heft twice their own body weight, it does not build endurance. The reverse is also true. A marathon runner, who logs hundreds of miles a week training his muscles to work painlessly for long periods of time, would never qualify for a Mr. America contest.

The striking difference in brute force and endurance between musclemen and marathon men lies in two distinct types of fiber that compose the body's 600 skeletal muscles. Specialized for stamina, "slow twitch" (type I) muscle fibers contract slowly but have a high threshold for pain and can perform work for long periods before they are exhausted. Built for strength and speed, "fast twitch" (type 2) fibers contract quickly but tire easily and in the process excrete lactic acid, which can accumulate in muscles, causing cramps and pain.

As muscle fibers are exercised, they adapt to the physical demands of different sports. When Swedish researchers examined muscle tissue extracted from the legs of ultramarathon runner Bertil Jarlaker before and after a 50-day jog across Scandinavia, they found that some of the type 2 fibers had been replaced by type I fibers, enabling him to complete the grueling 2,188-mile run. But the runner's amazing stamina exacted a price in strength and speed. Jarlaker's time per kilometer before the long run was about 4 minutes; after the run, he could do no better than 6 minutes per kilometer. The chronic inactivity of couch potatoes also causes muscles to make fundamental adaptations, in a negative direction, seriously diminishing both muscular strength and endurance.

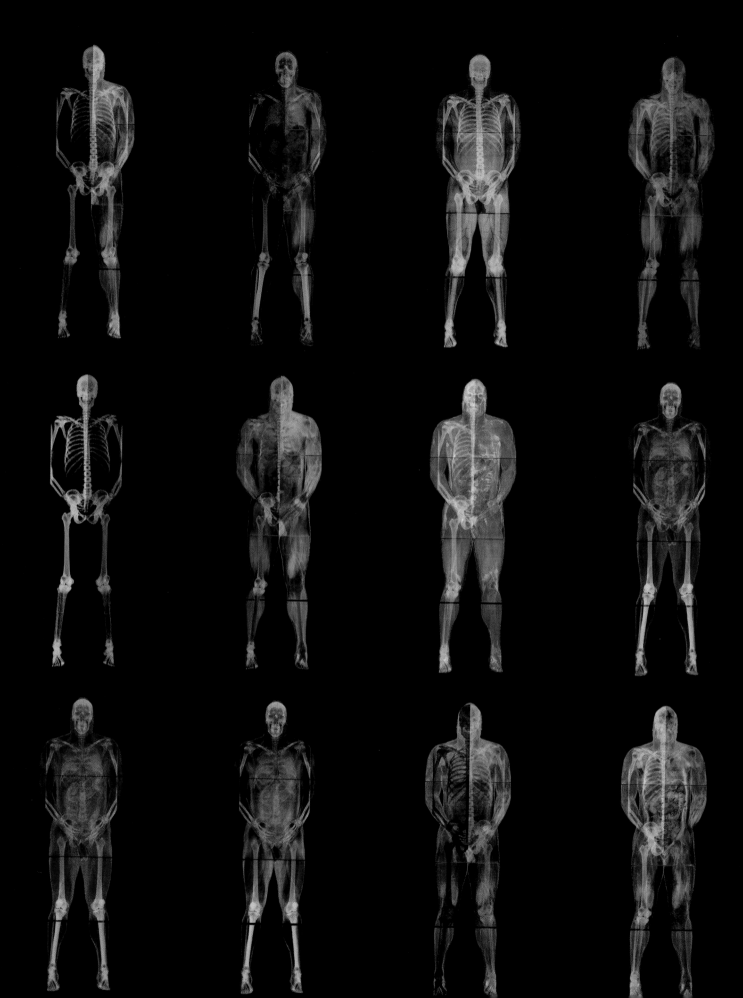

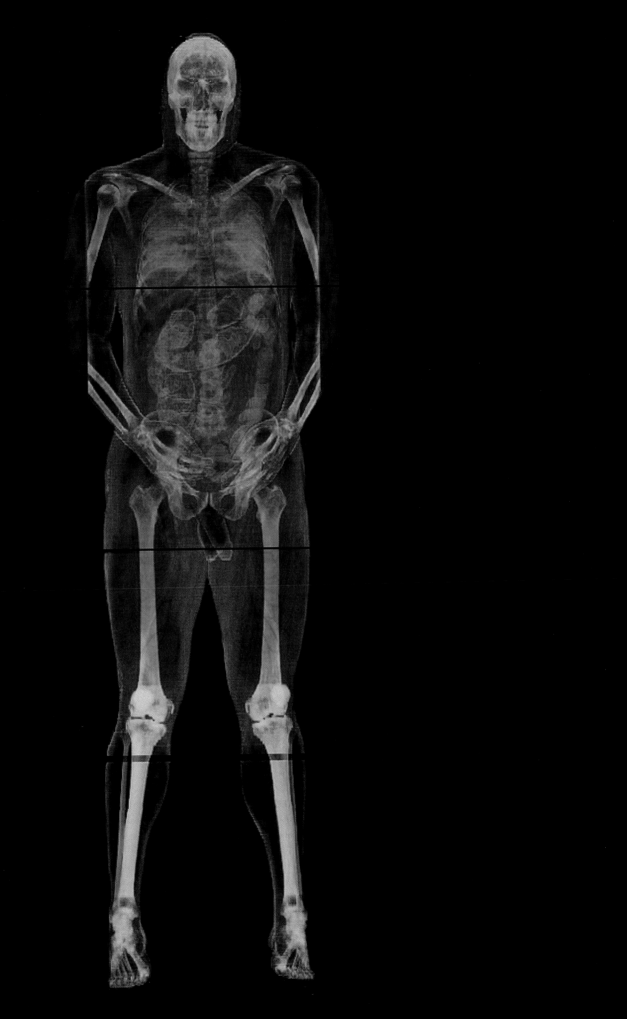

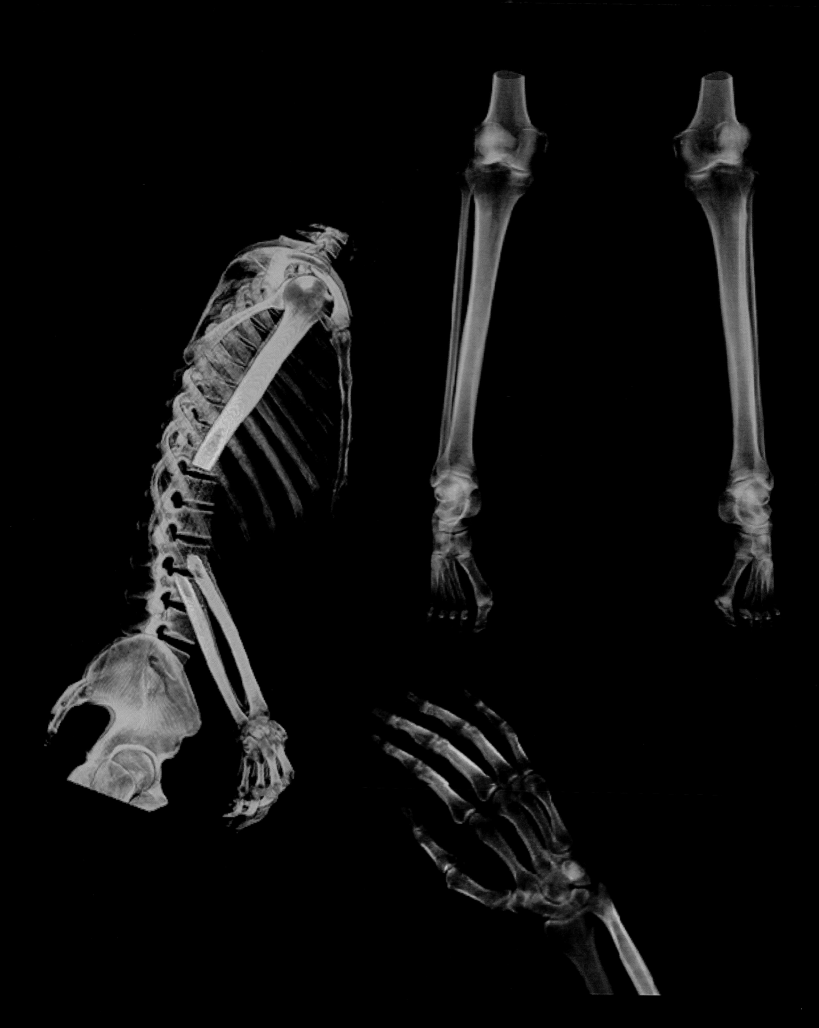

BUILDING REPLACEMENT BONES:

Like cellular anthills of perpetual activity, our bones are under constant construction. As demolition cells called osteoclasts destroy old bone tissue, carving out little cavities, builder cells called osteoblasts fill in the empty spaces with a cementlike mixture of collagen, calcium, and vitamin D, which hardens into new bone. The process, controlled by hormones from the parathyroid and thyroid glands at the base of the neck, maintains the strength and durability of the body's 206 bones, and as a result of their regenerative property, most broken bones heal readily. But when the production of bone slows down due to aging—both men and women over 40 lose a small percentage of bone mass each year—fractures can be disabling. Surgeons also find it hard to reconstruct complicated bones such as the jaw bone or large segments of bone that have been lost to disease or injury. Since bones not only regenerate but also remodel themselves into preset shapes, it is impossible, for example, to transplant a large bone segment from one part of the body to another without crippling the patient.

In search of alternatives, orthopedic surgeons have been testing a variety of bone-healing materials and technologies. They range from a quick-drying glue, which hardens into dahlite—a mineral that occurs naturally in the human skeleton—to electronic stimulators designed to boost the flow of minerals that turn on bone-building osteoclasts. Another unique solution to the problem of bone replacement—creating new bones from muscle tissue—has been tested in laboratory animals by plastic surgeons at Barnes Hospital in St. Louis. Working on rats, the researchers took flaps of thigh muscle and placed them in silicone rubber molds, which were implanted in the animal's abdomen to grow. The molds were coated with osteogenin, the chemical that signals fetal tissue to differentiate into bone and cartilage. Two weeks later the molds were opened, yielding perfectly contoured bone segments.

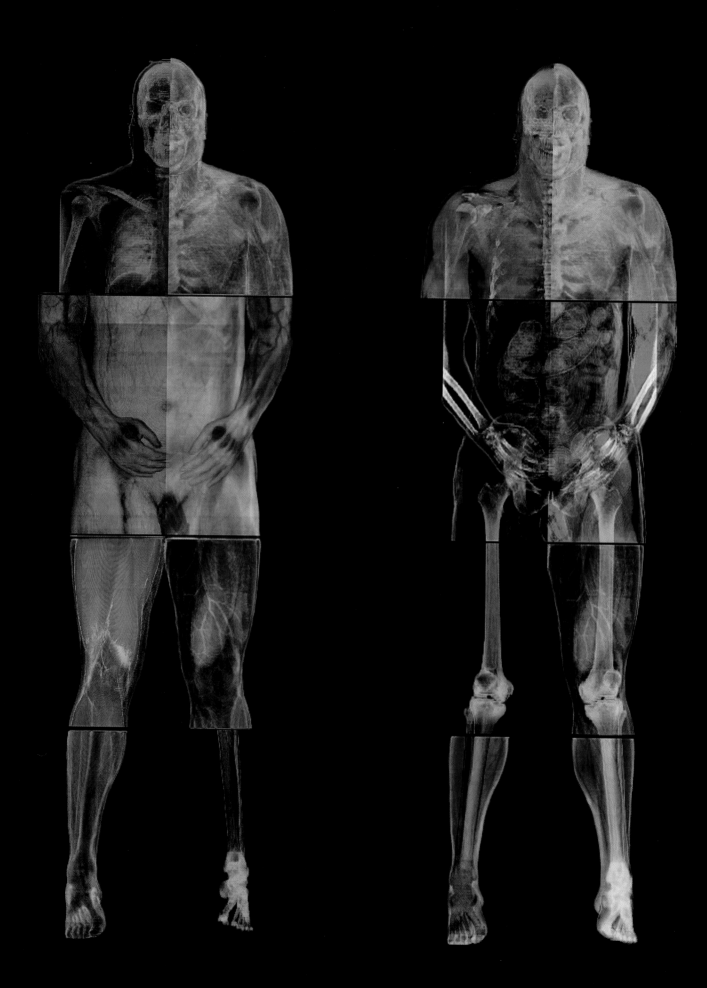

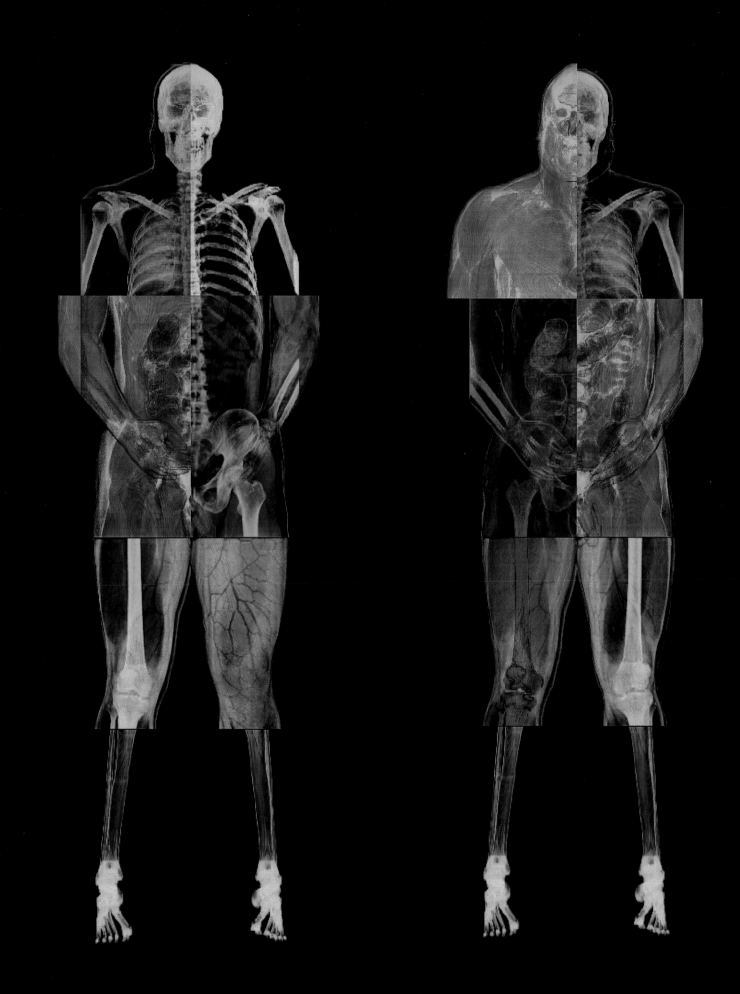

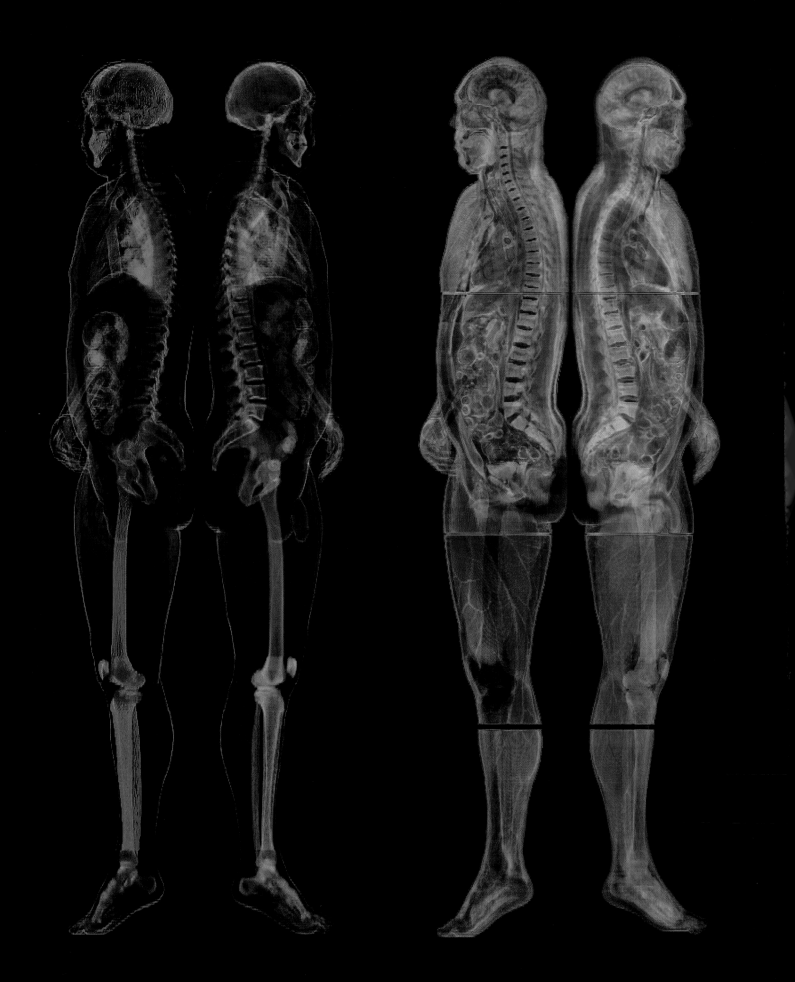

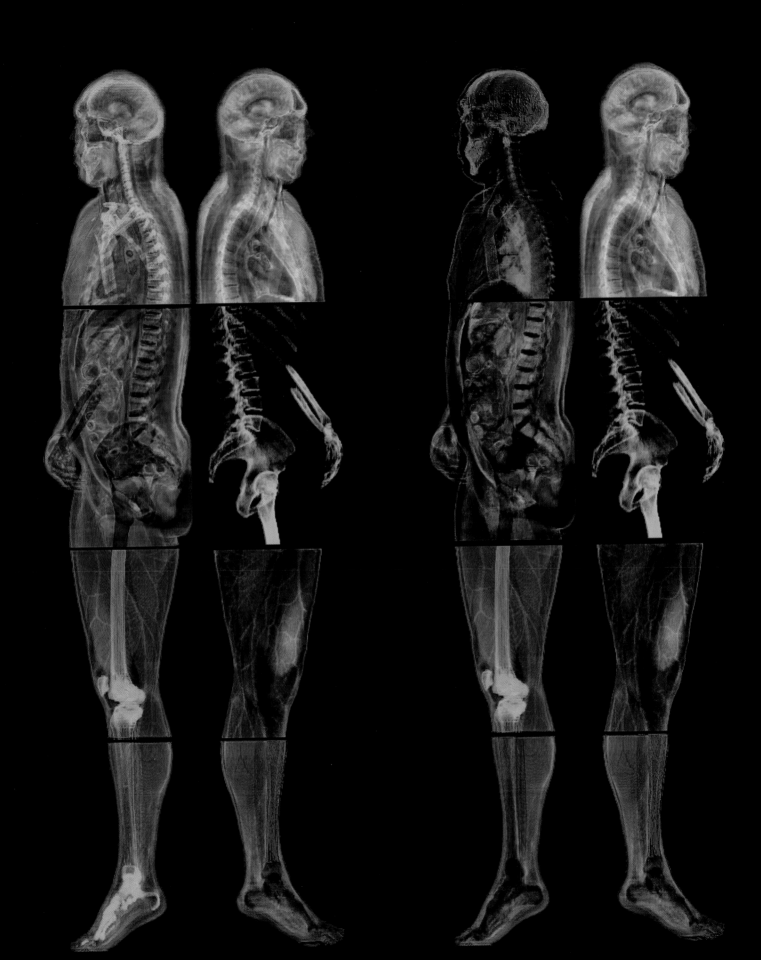

CLOSING THE GATES OF PAIN:

Stub your toe or bump an elbow and the almost instant reaction is to vigorously rub the sore spot. It's intuitive, everyone does it, and—most often—it works. Seemingly, this reflex is literally rubbing away the pain at the surface of the skin where the injury has occurred. But, in fact, the pain-killing effect really occurs in the spinal cord, where specialized cells, discovered by Ronald Melzack of McGill University and Patrick Wall of MIT, interpret pain signals coming from the body's periphery before these messages reach the brain. These "gate" cells, which shut down when pain signals become overwhelming, may play a vital self-preservation role in victims of gunshot wounds or other traumatic injuries, who often feel little or no pain. Rubbing a sore shoulder or scratching an insect bite can also short-circuit gate cells, by creating a flood of sensory signals, which compete with the original pain or itch messages. As a result, the molecular gates close, and pain disappears.

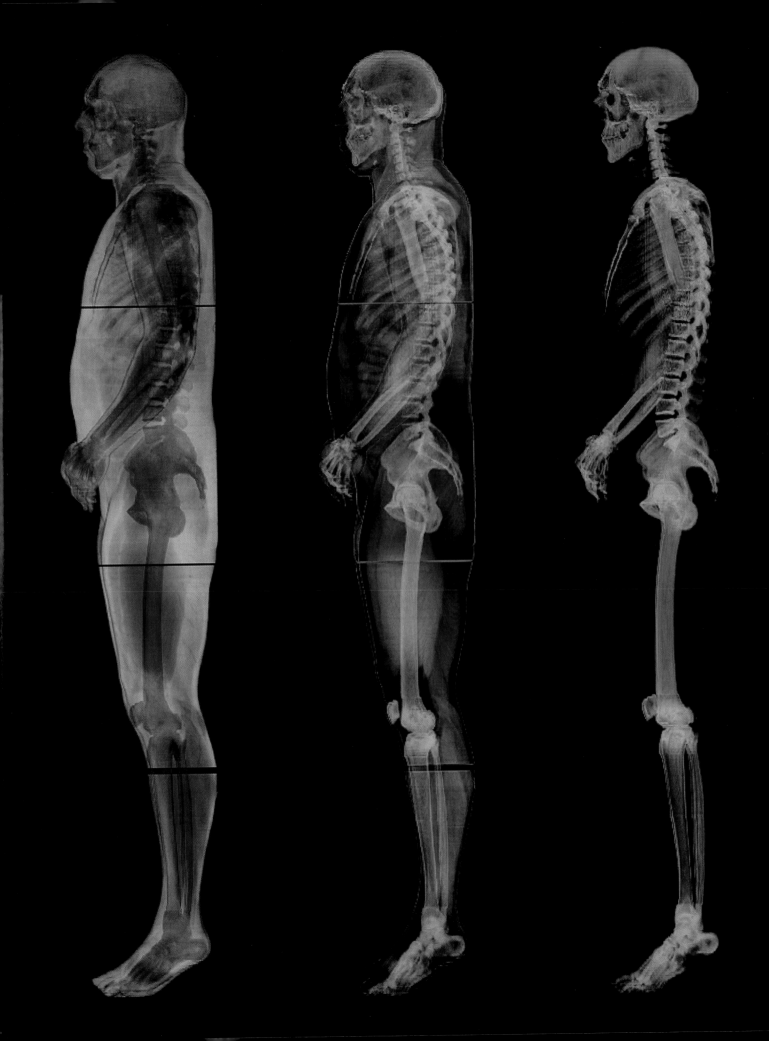

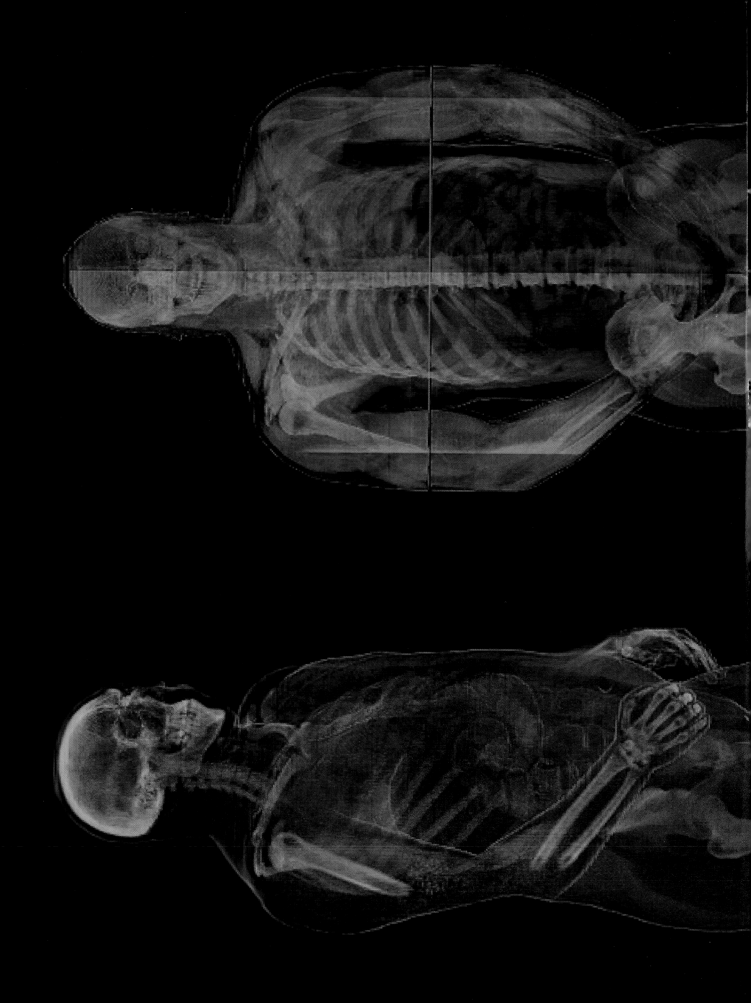

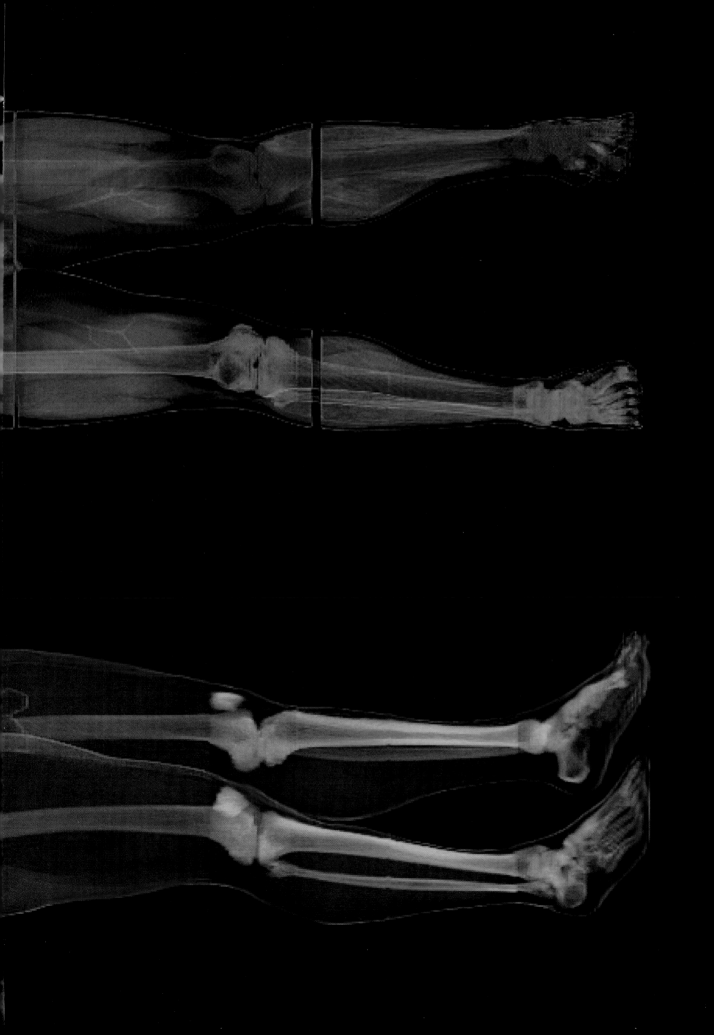